SpringerBriefs in Electrical and Computer Engineering

More information about this series at http://www.springer.com/series/10059

Syed Hassan Ahmed · Safdar Hussain Bouk
Dongkyun Kim

Content-Centric Networks

An Overview, Applications and Research
Challenges

 Springer

Syed Hassan Ahmed
School of Computer Science
 and Engineering
Kyungpook National University
Daegu
Republic of Korea

Dongkyun Kim
School of Computer Science
 and Engineering
Kyungpook National University
Daegu
Republic of Korea

Safdar Hussain Bouk
School of Computer Science
 and Engineering
Kyungpook National University
Daegu
Republic of Korea

ISSN 2191-8112   ISSN 2191-8120   (electronic)
SpringerBriefs in Electrical and Computer Engineering
ISBN 978-981-10-0064-5   ISBN 978-981-10-0066-9   (eBook)
DOI 10.1007/978-981-10-0066-9

Library of Congress Control Number: 2016931999

Printed on acid-free paper

This Springer imprint is published by SpringerNature
The registered company is Springer Science+Business Media Singapore Pte Ltd.

Preface

Initially, the Internet was invented to help people connect with each other at a very basic level. The maximum use of the Internet was to send and receive emails with some encoded text; in later stages, users were allowed to exchange limited-size images and textures with each other (known as the First Revolution in ICT). Later on, it started expanding to the new era of telecommunications, and we invented wireless phones. Similarly, we have witnessed Wi-Fi (802.11 family), which brought the Second Revolution in the world of telecommunications, where each user started to have an IP address. That IP address is used as an ID during communications. Furthermore, developers and various companies jumped in, and the socioeconomic competitive applications began. Moreover, news feeds—such as Yahoo News and MSN News—moved from being text based to video streaming in the form of live news channels. More precisely, the World Wide Web (WWW) enabled users to share a huge amount of "content" online.

Recently, it was stated that global IP traffic would be approximately 24 exabyte per month. This massive traffic is the result of the increasing number of users who access Internet on a daily basis to share large amounts of data. These statistics were collected by Cisco in its recent ICT forecast update report. Moreover, in 2019, it is expected that almost 4 billion users will be accessing the Internet using various mobile devices including tablets, cell phones, laptops, and smart wearables. In short, it is not a difficult assumption that the rapid increase in Internet use will bring several challenges to service providers. In the last decades, we have seen the demands of end users increase faster than research and development efforts in the area of telecommunications. For example, today we want to FaceTime, YouTube, and Skype on the go rather than use a simple landline audio call with fixed wires. Similarly, many essential software and systems require a significant amount of bandwidth. To be specific, the current Internet architecture was originally designed for fixed networking technologies along with the support of infrastructures. However, in the future, mobile devices will bring a new set of challenges, and these must be met in a new fashion. Therefore, we need to change the architecture of the Internet in the near future. These challenges called forth researchers, such as

Van Jacobson, to consider bringing about the third revolution in telecommunications, in which the host-centric focal point of communication becomes information-centric, and this is called an "information-centric network" (ICN). Since then, various architecture changes have been proposed by various thinkers around the globe. The main objective of this book is to describe those changes focusing on the known ones along with their possible applications in the future.

In this book, we first provide the overview of Internet communications and its varying and emerging architectures during the past decades. The purpose is to let our readers know the history of Internet technology. In the first introductory chapter, we discuss the history of the Internet and explain the reasoning to have new developments such as content distribution networks, peer-to-peer networks, and multi-cast communications. In addition, we perform a feasibility check of future Internet solutions presented in the literature as well as their requirements.

In Chap. 2, we discuss some serious efforts that have been made to bring about various architectures for the future Internet during recent years. Each of those architectures has one thing in common, i.e., to focus on content delivery rather than host-centric approaches. However, only few of them gained popularity due to their possible applications being investigated. Hence, we describe the overview of those varying future Internet architectures such as data-oriented networking architecture, content centric networking, named data networking, publish/subscribe, and network of information. The main objective of this chapter is to let our readers become familiar with the transformation of those architectures.

In Chap. 3 of this book, we focus on content-centric networks (CCN) proposed by Van Jacobson while working for the PARC Research Company. The main goal of the CCN was to change host-centric communication to content-centric communication. In CCN, the requester is known as a "consumer" who sends an "interest" to the network, and any node with the requested data can send back the "content" to the consumer by way of the same path. This simple overview seems superficial without explanation. Therefore, in this chapter, we provide readers with the history of CCN followed by its basic operations. Moreover, we describe the different components of CCN in detail as shown in the following examples show:

- What constitutes content? What is the structure of content and the interest message?
- How the interest can be forwarded and, in response, how can the data retrieval be efficient compared with the current Internet architecture?

We believe that this chapter will enable our readers to have a solid background in CCN, and in later stages, can become active researchers in the given field.

Due to them being in the early stages of development, content-centric networks (CCN) and their variants e.g., named data networks, are undergoing rigorous modifications to make them applicable to various future networks. Finally, in the fourth and final chapter of the book, we identify future research aspects and the issues that have been partially addressed or have not been properly tackled by the researchers. In addition, we also provide a road map for researchers in the relevant field.

The motivation for this book is to provide students and future information engineers with a useful collection of Internet standards, technologies, and techniques that derive from research projects for the management of distributed data. We wrote the book during a very exciting period of research advancements in the future Internet paradigm. We expect our readers to have a basic level knowledge of the Internet and wireless technologies.

Syed Hassan Ahmed
Safdar Hussain Bouk
Dongkyun Kim

Acknowledgments

This book would never be possible without the support and motivation of my beloved wife **Aiemun Hassan**. Her presence has meant a lot to me during the editing of my second brief book and research career so far. She has been my real inspiration and motivation for continuing to improve my knowledge and move forward in my career. She is my lifetime asset, and I dedicate this second book of mine to her as well.

Moreover, without the prayers and best wishes of my late mother, **Riaz Begum**, all of my achievements so far and forever would never be possible. Here my gratitude is due to my coeditor, **Dr. Safdar Hussain Bouk**, because he contributed to this book and without his motivations and dedications, it would have been impossible to achieve. Moreover, I am also thankful to my Ph.D. advisor and coeditor, **Prof. Dongkyun Kim**, who really motivated me and trusted me during the preparation of this book.

I also wish to acknowledge all the editorial staff at Springer, particularly **Annie Kang** and **Cindy Zitter**, who first helped us with the proposal to write this book and then maintained remarkable patience in light of our many delays in delivering the final files.

Syed Hassan Ahmed

First and foremost I acknowledge this effort to One Who said *"Can they who know and they who do not know be deemed equal?"* Only those endowed with understanding take heed.

Along with that, the love, affection, prayers, and motivation of my wife **Shaista Qaim Shah** and the beautiful smiles of **Mahpara**, **Imala**, **Jamana**, and **Hibba** were the main reason to complete this book.

I also acknowledge the prayers of my beloved mother **Amaan Dadi** and my father in heavens **Baba Bashir**, whose prayers always lead me to success. At last, I must acknowledge the push-and-pull strategy of **Syed Hassan Ahmed,** who drove me crazy during the course of writing this book. He is the one who started everything, and also he is the reason for my completing the text in time.

Safdar Hussain Bouk

This book would not have been possible without the valuable input, contributions, and feedback of my laboratory members Syed Hassan Ahmed (Ph.D. Student) and Safdar H. Bouk (Postdoctoral Fellow). I hereby acknowledge the extraordinary people I have had the honor of working with during my 15 years with the Wireless and Mobile Internet (MONET) Lab in the School of Computer Science and Engineering, Kyungpook National University, Korea. Here, I would also like to thank to my kids, family, colleagues, and friends, who always motivated and supported in pursuing my research and academic career.

This work was supported by Kyungpook National University Research Fund, 2015.

Dongkyun Kim

Contents

About the Authors

**Syed Hassan Ahmed** received his B.S in Computer Science from Kohat University of Science and Technology (KUST), Kohat, Pakistan. Later, he joined the School of Computer Science and Engineering, Kyungpook National University, Korea, where he completed his Masters and pursued his PhD in Computer Engineering at the Monet Lab. In 2015, he had been a visiting researcher at the Georgia Institute of Technology, Atlanta, USA. Since 2012 he had published over 50 international journals and conference papers in multiple topics of wireless communications. Along with several book chapters, he also authored two Springer brief books. He is also an active IEEE/ACM member who had been serving several reputed conferences/journals as a TPC and reviewer respectively. For three consecutive years, from 2014 to 2016, he won the best research contributor award for the workshops on future researches of Computer Science and Engineering, KNU, South Korea. His research interests include Sensor and Ad hoc Networks, Cyber Physical Systems, Vehicular Communications, and Future Internet.

Safdar Hussain Bouk was born in Larkana, Pakistan, in 1977. He received a B.S. degree in Computer Systems from Mehran University of Engineering and Technology, Jamshoro, Pakistan, in 2001 and M.S. and Ph.D. in Engineering from the Department of Information and Computer Science, Keio University, Yokohama, Japan in 2007 and 2010, respectively. Currently he is a working as a Postdoctoral Fellow at Kyungpook National University, Daegu, Korea. His research interests include wireless ad-hoc, sensor networks, underwater sensor networks, and information-centric networks.

Dongkyun Kim received a B.S. degree from the Department of Computer Engineering, Kyungpook National University, Daegu, Korea. He also received M.S. and Ph.D. degrees from the School of Computer Science and Engineering, Seoul National University, Seoul, Korea. He was a visiting researcher at the Georgia Institute of Technology, Atlanta, in 1999. He also performed a postdoctorate program in the Computer Engineering Department, University of California at Santa Cruz, in 2002. He has organized committees or technical program committees in many IEEE or ACM conferences. He received the Best Paper Award from the Korean Federation of Science and Technology Societies in 2002. He has been doing many editorial activities in several well-reputed international journals. Currently he is a professor in the School of Computer Science and Engineering, Kyungpook National University, Daegu, Korea. His current research interest includes connected cars, vehicular ad hoc networks, Internet of Things (M2M/D2D), Wi-Fi networks (including Wi-Fi direct), wireless mesh networks, wireless sensor networks, and future Internet. He is a member of the IEEE.

Acronyms

AH(s)	Authentication Header(s)
AVP	Attribute-value pairs
BLOB	Binary large objects
CCN	Content centric networks
CDN	Content distribution networks
CIB	Cache information base
CS	Content store
DONA	Data oriented network architecture
DOS	Denial of service
ERD	Enumeration request discovery
ESP	Encapsulated security payload
FIA	Future internet architecture
FIB	Forwarding information base
FTP	File transfer protocol
GENI	Global environment for network innovation
GMPLS	Generalized multiprotocol label switching
HTTP	Hyper text transfer protocol
ICN	Information centric networks
ICT	Information and communications technology
IETF	Internet engineering task force
IP	Internet protocol
IPSec	Internet protocol security
LCD	Leave copy down
LCE	Leave copy everywhere
LFU	Least frequently used
LPM	Longest prefix match
LRU	Least recently used
MAC	Medium access protocol
MANETs	Mobile ad hoc networks
Message TLV	Message type, length, and value
MPC	Most popular caching

MTU	Maximum transmission unit
NDN	Named data networks
NFC	Near field communication
NSF	National Science Foundation
P2P	Peer to peer networks
PARC	Palo Alto Research Center
PIT	Pending interest table
Pub/Sub	Publish and subscribe networks
QoS	Quality of service
QR	Quick response
RFC	Request for comments
RID	Regular interest discovery
RSS	Really simple syndication
SDN	Software defined networks
TCP	Transfer control protocol
TRIAD	Translating relaying internet architecture integrating active directories
URL	Uniform resource locator
VANETs	Vehicular ad hoc networks

Chapter 1
Introduction

Syed Hassan Ahmed, Safdar Hussain Bouk and Dongkyun Kim

Abstract During the 19th century, ICT services and innovations have enabled humans and machines to interact in various ways. The very basic and most important invention was the "Internet," which actually connected and thus became the baseline for the rest of the upcoming discoveries in almost every scientific field of research. Today, the Internet is used by everyone regardless of their location, and people are depending on the Internet more than ever expected, even by its initial developers. In fact, it is not a difficult assumption to say that the rapid increase in the use of the Internet will bring several challenges to service providers. lasting past decades, we have seen the demands of end users increase faster than research and development efforts in the area of telecommunications. For example, today we want to FaceTime, YouTube, and Skype on the go rather than make a simple landline audio call with fixed wires. Similarly, much essential software and many systems require a significant amount of bandwidth. In short, we need to change the architecture of the Internet in the near future, and we expect that the new architecture will contribute to change the focal point of communications from host-centric to information-centric because today we are interested in services, rather than sources, providing content. In this introductory chapter, we will discuss the history of the Internet and why we require new developments such as content-distribution networks, peer-to-peer networks, and multi-cast communications. In addition, we enlighten our readers with the possible outcomes of the current research going on in the field of the future Internet.

Keywords Internet history · CDN · P2P · Future internet · Communication challenges

1.1 Introduction

Initially designed for secure yet limited use, such as e-mail exchange and file transfer, the **Internet** architecture has been transformed from the past decades. Later on, connected devices started sharing database registers, activating printers, and accessing files from online servers and so on. Therefore, we have witnessed

© The Author(s) 2016
S.H. Ahmed et al., *Content-Centric Networks*, SpringerBriefs in Electrical and Computer Engineering, DOI 10.1007/978-981-10-0066-9_1

enormous research and development efforts being made for improving communication among hosts, clients, and end users. Hence, we believe the fact that the Internet emerged as an end-to-end communication network for sharing data and resources between devices [1].

With the passage of time and amazing growth during the technological era, the theme of communications has shifted from basic-level communication to content dissemination [2]. Here it is worth mentioning that the content can be a text file, an image, a short video, or larger data traffic. In contrast, new technological trends tend toward the need for increased bandwidth. No doubt increased bandwidth brought us a whole new set of applications including multimedia applications [3]. Moreover, in our daily life, the pervasive access of the Internet has presented a complete new set of businesses—such as online marketing, search engines, social networking, advertisements, and online commerce—that have been built on the aggregation and sharing of personal data. For instance, video-sharing websites (such as YouTube, Daily Motion, etc.) and peer-to-peer (P2P) networks are trending examples. Such applications strengthen our observations that content distribution over the Internet has grown from a textual toward a multimedia information system where services and applications focus on content [4].

This rapid growth in the volume or size of the data being exchanged over the Internet brings up several questions as follows:

- Is the current Internet going to be able to successfully handle content-centric communications?
- How can mobility be handled while having the same IP-based architecture?
- What security protocols will be sufficient to provide security for the massive amount of data?
- Will current infrastructure-based mobile networks be the only solution to handling mobility?

In short, there are many questions regarding the feasibility of the current Internet architecture, designed decades ago, to support our communications in future [5]. In history, we can see developments and advancements in the result of the enormous efforts initialized by researchers from both industry and academia. However, most of the advancements have been to improve IP-based networks and the discussion regarding how to address the contents instead of the end-to-end hosts remains very much alive.

1.1.1 History of the Internet

As mentioned previously, during the initial days Internet applications were mostly based on word-based information only. Similarly, users were expected to use the Internet for exchanging email messages and transferring files by way of file transfer protocol (FTP) and access remote servers. In contrast, today the Internet is a complex and heavily loaded multimedia/information system based on content distribution [6]. For simplicity, we call documents, videos, audio, images, and web pages the "contents." Similarly, the metadata used to find, understand, and manage such contents is also included in the heading of "contents." Therefore, the new system must enable users to request and then receive the required content in an efficient manner [7].

For that purpose, first, content resolution is required, and thus it should be guaranteed by the system we design. "Resolution" here means that there should be a unique identifier of any content, and this can be achieved only when there is a state-of-the-art mechanism available to generate those identifiers. In addition, the lifetime of the content should also be considered as a parameter while setting the identifier. For example, let us consider the terms "chronological" and "perpetual." Data or content can be chronological if it has a shorter lifetime or validity such as a weather report, a road traffic condition, an emergency situation, any error in any system, and so on [8]. Perpetual or long-lasting valid data include location information on a fixed server of a building, a street, a data rate, the capacity of any system, and so on. Recently, it has been perceived that the advent of Web 2.0 has increased the number of content publishers. This means that users with no IT background and technical soundness can publish or upload the content of any size over the Internet. That content is most likely to be requested by users later. Hence, we can say that it is nearly impossible to assure a complete persistence of content while taking the current Internet architecture into account.

Second, it is expected that the new system must be scalable in such a fashion that the search and forwarding mechanisms of the contents are efficient regardless of the total number of users, their users' locations, and the content offered. Both the content users and providers must be able to operate at the Internet scale.

Finally, secure access to contents is key to providing authentication and further access-control mechanisms to the contents available. So far, there is no such solution or system that satisfies all the aforementioned requirements at the same time. However, the current literature is full of investigations being made to partially satisfy them. Figure 1.1 shows the revolutionary breakthroughs made so far in the IP-based Internet paradigm. It has been rationalized that the currently the communication focal point is content rather than the devices. Hereafter, this chapter will focus on the communication perspective on the current Internet and its limitations for future requirements.

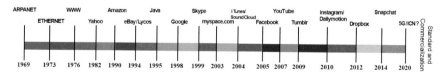

Fig. 1.1 Revolutionary advancements in Internet history

1.2 Communications in the Internet

In this section, we focus on the basics of Internet communications. Readers are provided with historical evolutions in Internet communications varying from end-to-end concepts to the latest applications of publish/subscribe systems.

1.2.1 End-to-End Communications

On a worldwide scale, the Internet is a packet-switched network where packet forwarding is based on the maximum best-effort service module possible by the Internet protocol (IP) [9]. During the process of packet forwarding, however, neither a resource nor any different predetermined service has been determined. As a result, contents are distributed without taking performance into account. Similarly, the current Internet architecture focuses on communication between hosts. This means that the name of the host as a source is included in the packet header in the form of an IP address, and the header also includes the IP header of the destination host. During the communication, the packets are traversed hop-by-hop, and this traversing is solely based on the destination IP address. Initially, this model was suitable for the Internet applications of that time where the main goal was to share resources remotely offered by a particular host, such as a printer server, a file server, and a Web server [10]. However, in the current era, it is difficult to satisfy the content distribution and its requirements due to various constraints such as the location information of the contents and the name under which the content is stored on the destination server. For this reason, content distribution in the current Internet is supported by patches, which consist of a set of mechanisms and protocols, thus partially satisfying the application requirements accordingly.

One known Internet protocol is the hypertext transfer protocol (HTTP), which redirects the content search, especially those with a nonpersistent nature [11]. Therefore, HTTP objects are usually requested using the resource locators and are referred to as "uniform resource locators" (URLs). The relevant URLs are included in the headers of the packets or messages being sent. Through this process, the HTTP forwards the events triggered by the server hosting the requested object or contents. In this case, an HTTP [12] redirect message is sent back to the client that contains the new URL in its header field. This process, however, must be aware of the content's location, and hence it is mandatory to have a complimentary mechanism assisting the persistent access to the required or requested contents irrespective of their location, characteristics, and other properties. Our example shows the working model of client server model. For instance, one point-to-point communication channel is built between one client and one server. Let's assume that several users simultaneously send a request message for a given content located at or hosted by the server where multiple point-to-point channels are established and single copy of content is to be sent over each channel. Hence, we conclude that the most popular content results in less efficient distribution, specifically in terms of bandwidth. Despite the fact that it is not efficient, this model is being widely adapted by the current content distribution applications and architectures [13]. In short, the content distribution for applications on a large scale requires significant improvements in forwarding mechanisms to make them more scalable and different from the current client server model(s). In addition, the authentication and security of the content cannot be avoided; hence, it is expected that the applications for content distribution also provide security for the content during the Internet communication. However, in the current Internet system, security has been guaranteed for the channel being used between the source and destination host(s) while ignoring the security of the content(s).

Therefore, for the sake of security, a few additional processes and messages have been introduced, for example, Internet protocol security (IPSec) [14]. The IPSec causes overhead by introducing a patch that is used to ensure secure connection/communication. To be specific, IPSec enables users to maintain reliable and secure connections with the help of authentication headers (AHs); in addition, cryptography is applied to the data by encapsulated security payloads (ESPs) [15] and various key management mechanisms. Likewise, the content security still depends on the host and its relevant channel security levels. Hence, the scalability is still an open issue and requires appropriate attention from the research community [16]. Because the same content is not shared between users due to the unsecure channel between two or more hosts, an alternative method is to establish multiple secure connections among the content users and sources, which is of course going to increase the number of channels and security issues if the number of users increases. Therefore, we believe that some specific solutions for content distribution applications are still missing.

1.2.2 Multi-cast Communications

Here we discuss the multi-cast communication process, which is one of the initiatives for making effective content distribution possible over the Internet [17]. During development, the multi-cast is implemented by integrating an IP multi-cast on top of a medium-access control (MAC) layer. To be specific, in multi-cast communications, a single datagram of the content forwarded by a host may reach multiple destination hosts interested in the content. Basically, these destination hosts are accumulated in the form of group, which is originally identified by a single IP address. That's why the source host needs to send one datagram to that one IP address. All of the connected hosts to that one destination IP address receive the datagram/data accordingly. In this case, the network layer plays a vital role in forwarding and replicating the datagrams, sometimes, over the distribution tree, thus covering every host interested within that group. The main advantage of such approach is to avoid unnecessary copies of the same datagrams over the link during the communication process. However, we must know that IP multi-cast was originally proposed in the 1990s and has not been adopted on the larger-scale Internet [18]. According to several researchers, one reason can be complexity of the system to configure, manage, and allocate the set of protocols required by the IP multi-cast. In short, a host is able to join and leave the group any time, and also it might be a member of more than one group at the same time. Moreover, it has also been reported that the sender is not required to be a member of the group. For instance, this approach leads us to the issues of authenticity and privacy.

1.2.3   Peer-to-Peer Systems

In the recent past, researchers from academia and industry developed a great idea of letting users share content with each other regardless of their location and other factors. Servers, instead of providers, started playing the role of the "bridge." However, the basic infrastructure is still IP based. The founders of that time named this new emerging technology "peer-to-peer" (P2P) systems [19]. Content sharing in P2P takes place in such a fashion that nodes with similar interests create an overlay network at the application layer and are known as "peers." These peers moderately share the bandwidth, the downloading process, and the storage capacity over the servers, thus resulting in efficient content retrieval. The basic idea is that the peer contributes to the given amount or the limit of its resources and uses the services made available by the overall P2P system [20]. As a result, additional peers in the systems contribute to the efficiency of the content available. The scalability of P2P system, however, still depends on the number of peers involved. Moreover, the P2P system is independent of any changes to be made in the network core compared with multi-cast IP systems [21].

In today's era, the user is interested in retrieving the required content regardless of the source. For example, in the current communication system, we can find in BitTorrent, where every new peer in the system chooses its partners or collaborator randomly and start collecting chunks of the required data. These partners were selected from the group of the peers already registered, or they have the same interests of the required content, and here we must clear that the location or identity of the peers involved is not mandatory or being used. The massive success of P2P networks for both file sharing and streaming applications has proven the fact that the Internet paradigm must be changed [22]. Here we argue that P2P is the baseline and runway for information-centric networks (ICNs) to take off.

It is expected that ICNs will be more scalable than P2P. Although P2P is providing good throughput and more efficient content retrieval [23], it suffers from security problems; in addition, there is a limited source of incentives for peers to share their resources. Moreover, P2P networks rely on peer collaboration to work properly [24]. Hence, whether or not the data forwarded by the other peers are trusted is a critical question to be answered by future networking paradigms such as ICNs. In addition, the robustness of the system is directly related to the number of peers joining or leaving the network. In the current Internet architecture, there is no dedicated architecture or infrastructure to deal with (1) the management of those peers joining or leaving and (2) determining the remaining chunks out of the total chunks for any content.

1.2.4 Content-Distribution Networks (CDN)

Another attempt to increase the efficiency and scalability of the client server communication model has been made by proposing content-distribution networks (CDNs) [25]. No doubt CDNs are improving many applications for content distribution. Basically, CDNs consist of a group of distributed systems, and that group is interconnected by way of the Internet. Those systems cooperatively contribute in content distribution. To be specific, the contents are replicated on various servers, mostly by different Internet service providers (ISPs). This feature enables CDNs to increase the availability of any content. When a user requests any content, x, the request for x is forwarded and redirected to one of the servers close to the user. By this, CDNs try to minimize hops between the requester (user) and the provider (source server) and thus decrease the latency and increase the delivery rate due to the probability of less congestion [26]. CDNs are comprised of two main building blocks, i.e., the replication and distribution service and the request redirection service, where the content providers (servers) use the former service to find proper servers and to allocate storage capacity and so on. In contrast, the latter service is used as an interface between content consumers and providers.

Fundamentally, this service assists in receiving requests for the required content and later forwards each request to the most appropriate CDN server in order to satisfy it. Moreover, CDNs are typically composed of two types of servers: an origin server and a replica server. The origin server attributes the content identifier and is responsible for storing and announcing the contents. In contrast, the replica server forwards the content to the clients [27]. Generally, clients send request(s) to the origin server, which redirects these messages to the replica server closest to the client that stores the desired content. Figure 1.1 illustrates this process. In summary, redirection mechanisms severely affect a CDNs performance (Fig. 1.2).

1.2.5 Publish/Subscribe Systems

Here we give one example that shows that the current Internet architecture must be changed in the near future. That example is the recently proposed Publish/ subscribe architecture, which is also known as pub/sub [28]. The pub/sub system also supports quite an identical mechanism to that of P2P for content retrieval. In pub/sub, the users are interested in receiving the content(s) only regardless of the identity of the sender. For that purpose, the contents are named "events," and

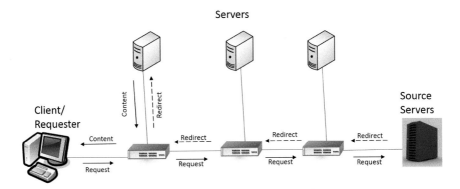

Fig. 1.2 Basic operations and scenario of CDN: A client sends a content request to the source server(s), which redirects this request to the server(s) closest to the client. Then the closest router sends the content to the client

their delivery is known as "notifications." The users basically subscribe for an event (i.e., content/data), and later on when the provider node or server finds that event in its repository, it replies back with the data, and that reply is supposed to be in the form of a notification. Here we describe the basic operations of pub/sub as follows:

Initially, the publisher node(s) create(s) event(s) and make a list of those events available to the subscriber(s). Later on, the subscriber nodes are able to initiate their interests in various events and patterns of events under the given preferences by the actual publisher, due to which subscribers are notified whenever is an event matching their interests is propagated by the publisher. There is a possibility that one publisher announced a list of events and got subscriptions; however, due to the factor of mobility (if applicable), the actual publisher moved away from the subscribers. In this case, other publisher for same events can be notified using some backbone networks. Here we must note that the subscribers and publishers are decoupled both in terms of time and space. As mentioned previously, publishers may announce their interest in any event, which is not yet published. In addition, the interest is not necessarily to be announced when the publisher of the associated event is online or in transmission range. Therefore, decoupling in this context somehow guarantees the scalability of the pub/sub system. In addition, it allows publishers and subscribers to work independently (readers can refer to Fig. 1.3).

The beauty of pub/sub is that it supports content distribution between a noticeably huge number of users because the publishers do not store information related to the interests of the subscribers, and similarly subscribers can receive content from any publisher regardless of whether the sender is known or not. The first pub/sub was proposed on the basis of a topics subscription. It allowed users to subscribe to a specific topic such as stock exchange information, weather reports for a specific area, and so on. Really simple syndication (RSS) feeds [29] are one of the successful milestones achieved, and they allow researchers to look into the topic later on. In topic-based pub/sub systems, users subscribe to events by using a topic

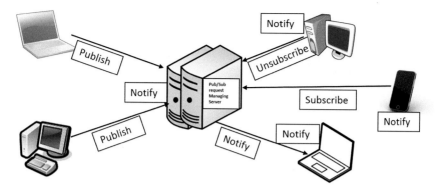

Fig. 1.3 Subscribe and event-notification functions in a simple pub/sub system

as a keyword, and whenever the publishers obtain fresh data on that event, the users are notified. Bloggers, Web sites, news channels, and any other applications have adopted RSS. The basic operations of pub/sub are quite similar to the concepts of IP multi-cast as described previously in this chapter. Each topic is a unique event, identified by the unique name, and must provide users with interfaces to use subscribe and publish functions.

Content-based systems are quite relevant and will be the next step in the evolution of pub/sub systems. In content-based systems, users subscribe to events based on the characteristics of the events themselves rather than on previously defined and static properties such as the identity of the topic. By this, the users are able to filter their subscriptions by using restrictions based on attribute value pairs (AVPs) and other basic logical operators, for example, $<$, $>$, $=$, and etc. Moreover, using Boolean operators and other looping techniques, the restrictions and conditions can be combined.

To be more precise, the various architectures of pub/sub systems are classified into centralized and distributed ones. The manner of subscription does not affect these classifications. Centralized pub/sub systems allow publishers to send messages to the central unit, e.g., a server, which stores all of those messages and redirects them to the subscribers when demanded. In contrast, the distributed architecture allows all system nodes to forward interests/notifications and process the requests because there is no central entity to take care of these various operations. Both strategies have their own merits and demerits. The authors here assume that the readers must have some preliminary knowledge of distributed and centralized architectures for different wireless and wired networks [30]. Generally, the distributed systems rely on multi-cast communications and are therefore prone to deliver the content more efficiently. Topic-based pub/sub systems offer benefits in terms of distributive properties. In contrast, content-based systems usually face many challenges if the distributed architecture is used due to the multi-cast communications. Multi-cast communication is influenced by the computational cost of filtering the subscriptions and forwarding the content.

1.3   Communications Challenges

No doubt, the Internet has changed our lifestyles in a positive manner. In addition, the contributions of the Internet to the growth of business cannot be downplayed. Moreover, the Internet has been playing a vital role in improving national defense and other important areas including hospitals, transportation systems, and economics. However, the fact is that most of the Internet architecture perspectives were designed and implemented almost 30 + years ago. In those 30 years, there has been extensive research and development about the networking and packet-switching modules. One question here arises: Is the current architecture appropriate way if we were given a chance to start the Internet today? In this chapter, we aim to answer this question, which has been raised by the National Science Foundation (NSF) while designing the next-generation Internet architecture known as the Global Environment for Network Innovation (GENI) [31]. More specifically, the next generation of the Internet is expected to be friendly for commercial use. For simplicity, we name the next generation of the Internet as "Internet 3.0." The reason behind this is to make sure that we are talking about the latest version compared with Web 2.0. Moreover, the NSF is looking forward to continuing this program of the next-generation Internet in line with the GENI initiative. According to the recent reports, the NSF plans to invest millions of dollars in this project, thud making it one of the largest projects the NSF has ever invested in. Hence, in the coming years, more research centers and academic giants will be attracted to and perform research for this project. Here we discuss the overview of the best available sources extracted from those existing works while putting them together in a coherent interoperable way.

Here it is worth stating that the Internet is a 40 -year-old technology because the first Internet RFC is dated approximately April 1969. So far, many steps have been taken to improve the Internet. In fact, we have witnessed the borrowing of ideas from current Internet-working research as well as from the other networking domains such as telephone, airlines, highways, railroads, postal services, and walkways. We came up with the solution that the next-generation Internet should be more secure compared with the existing version. In addition, it also should redefine the boundaries of the business as well as the policies inside those boundaries. The enormous use of Internet missionary governments sets further rules to protect citizens from cyber-crime in the same way that the government is responsible for protecting citizens from other hazards. Moreover, a new set of rules is required, and those rules should be flexible so that different governments can set various rules on top of the general ones. More precisely, the future Internet architecture should enable users to design mobile objects. For instance, we have new literature for software-defined networks (SDNs) that enables users to virtually set network functions according to their requirements [32]. However, those configurations require specific levels of training and infrastructure so that companies can handle them. We are expecting those complex operations to be turned into simple ones for end users. This new architecture can be based on naming content and allow users to decide from where they would like to receive Internet traffic, control the privacy of their location, and so on.

1.4 Future Internet Technologies: A Solution?

From the previous discussions, we are convinced that we must move the current Internet architecture toward the emergence of new technologies collectively known as the "future Internet." In the future Internet, we have a rich literature available for ICNs [33]. The current Internet relies on purely point-to-point primitives, whereas according to the definitions of ICN projects, we are able to move the new Internet toward the data- or content- centric and -oriented networks. Almost a decade ago, a TRIAD paper proposed an architecture similar to ICN, and hence it should be given the credit to allow researchers to start working on future Internet architectures such as ICN. Later on, Baccala wrote an IETF draft in 2002, right after TRIAD, that also indicated that we should move forward from the primitive operation of displaying a Web content by way of end-to-end connection toward the delivery of a named set of data [34]. Moreover, few more researchers during those years claimed that they were ahead of the mentioned work. However, remarkably vatic, both of the designs still used the existing DNS-naming schemes, along with its inherent drawbacks, and only focused on basic content delivery with minor attention paid to other important issues including, but not limited to, security, streaming media, and faulty servers [35].

In this context, the data-oriented network architecture (DONA) was proposed almost 5 years after the former two identical works [36]. DONA was the first comprehensive and detailed clean-slate ICN design that supported the use of self-certified content names suggested by earlier works of its time. Different from the others, DONA also incorporated advanced cache functionalities to address the various ICN issues. Unfortunately, rapid follow-up on the work appeared later with reference to other ICN architectures, and this caused avoidance of the research topic to be pursued by the broader research community later.

Recently, content-centric networks (CCN) have been proposed with the ignition of interest in the ICN area, and the idea has spread widely [37]. The CCN inventor Van Jacobson took on this initiative while working for the Palo Alto Research Center (PARC). Several workshops have been devoted to CCN and ICN, and projects such as 4WARD, PSIRP/PURSUIT, SAIL, and COMET have been focusing on the given topic. Later on CCN became the preliminary architecture for the Named Data Networking (NDN) project [38], which was chosen as one of the four proposals in the Future Internet Architecture (FIA) program lead by the NSF. However, ICN research is still difficult to be adapted into real-time scenarios. There is little common terminology between these proposals, and because there is no common or standard framework yet, the focus is often on low-level mechanisms being performed in the recent research. Due to this early stage, many research works emphasize the differences between design and others while leaving the readers to design/construct the "forest" of ICN out of these proposed tree-like structures. In contrast, the research community had taken it for granted so far, but the reality is that ICN deserves to be explored rapidly and widely. These current affairs led us to compile this book with the goal of providing a broader perspective on current ICN designs and their history to give readers a complete set of knowledge, and thus motivation toward research activities can be increased.

1.4.1 Fundamental Differences in Design

Now we will discuss several aspects of ICN that make it different from the current Internet. Mainly, ICN differs from the current Internet architecture in three important perspectives. Those include (1) naming the content, (2) interdomain routing, and (3) the location of the narrow waist with an ICN-based Internet. As discussed previously, in ICN each client or user looks up the content by its name while being unaware of its location. Similarly, it is expected that any content-oriented security model must hold the following properties:

  (i)   The consumer is expected to know the exact name of the content and the type it is looking for. That is, the consumer must be able to map between the real-world description of what they require (e.g., BBC headlines) and its corresponding ICN name resolved by the system.
  (ii) The consumer should know the provider's public key for any content so that it can verify the attribution and integrity of the retrieved content.
 (iii) The ICN system as a whole should be able to allow binding of an object's name to the public key of the content provider so that it can prevent attackers from registering false content(s). Avoiding this binding may result in favors of attackers who may use false content as a denial-of-service (DOS) attack.

In terms of naming, ICN consists of two main naming systems. The first one resembles today's DNS names and uses hierarchical human-readable names [39]. The human readability partially addresses the first requirement, and the hierarchical structure helps with scalability. There can be multiple ways or techniques that may allow the requester/consumer to know the public key (ranging from personal contacts to webs-of-trust to PKIs), but for the ICN system to be aware of this key requires a globally agreed-upon PKI to bind names to keys. In contrast, the second naming system takes self-certifying names into the account. In this case, the public key is bound to the name itself, so the ICN system does not depend on any PKI. However, these names are not human-readable, so requesters must use some other techniques (e.g., search engines, personal contacts, webs-of-trust) to determine the name of the required content.

For the satisfaction of interest packets being generated for any content, the ICN systems must route those requests or interest packets. To make this routing happen, there are many different approaches in the ICN literature for achieving name-based routing within a given domain, but these differences are largely autonomous in nature and not of fundamental importance. Hence, the ICN proposals differ more fundamentally in terms of routing, i.e., how the routing is performed between domains. For example, a CCN leverages the current interdomain routing system and builds name-based routing on top of BGP. Here it is to be noted that CCN is one of the ICN architectures. Others (such as DONA) follow the BGP policy model, but they do their own name-based routing, and still others (such as PSIRP) develop their own interdomain routing paradigm.

Likewise, as we know, IP is the narrow waist of the current Internet architecture. This is frequently seen as a major difference between ICN designs, but we stand firm that this is actually a broader architectural debate that is mostly orthogonal to ICN design details. All of the ICN designs involve hop-by-hop communication between the ICN-layer elements (e.g., content routers in CCN, rendezvous nodes in PSIRP, resolution handlers in DONA, and content-aware routers in curling). Because this communication is merely between hops (and does not require global reachability), one could carry any of these designs over IP or as a substitute for the IP (running over some L2-like layer that offers local delivery). Nevertheless, whether one retains or replaces IP as the narrow waist obviously has implications for the overall Internet architecture as well as for the performance required in the ICN layer [40]. More detailed analysis and descriptions will be presented in Chap. 2 and 3 of this book.

1.4.2   Anticipated Changes in Future Internet Technologies

In this section, we list the top 10 features that would help remove some of the problems faced by current Internet users.

(i)   Energy-efficient communication: Current Internet architecture requires both source and destination end-systems to be up and awake for the communication to take office. All packets received when the destination is down are dropped. By introducing wireless devices, this restriction is rendered tranquil by allowing base stations to store the packets while the subscriber device is sleeping. For energy-efficient communication, this should be generalized to wired devices as easily.

(ii)   Separation of identity and address: In the current Internet, a system is identified by its IP address. As a result, when a system changes its point of attachment, the address changes. This makes reaching mobile systems more difficult. We agree with the fact that this is a well-known problem, and a number of attempts and proposals have been made in the past to solve such issues including mobile IP, internet indirection infrastructure, host-identity protocol, and others.

(iii)   Location awareness: IP addresses are not linked to geographical position. This can be considered a strength of IP. However, a large share of information-transfer applications, such as any other transport system, requires finding the nearest server. Likewise, mobile nodes must know their position. The next-generation Internet should have the receiver decide about location privacy.

(iv) Explicit support for client server traffic and distributed services: A large percentage of current Internet traffic is client server traffic. A Web user trying to reach Google is a model of client server traffic. These users tend to reach "Google," which is not a single system. In fact, it is a distributed service with hundreds of systems in hundreds of various locations. However, the user is interested in communicating with the nearest instance of this service for the quickest and most relevant response. In the current Internet, the name "Google" resolves to a single IP address, and so directing users to the right server is unnecessarily complex.

(v) Person-to-person communication: The Internet was originally designed for computer communications. However, the real target of communication is often a human being. In today's world, a human being may be reachable using a desktop computer, a laptop, a cell phone, or a landline (i.e., wired) phone. The main goal is to reach the person and not the desktop computer, the laptop, or the phones, respectively. Here, we recall in the section on IP addressing that the person does not have an IP address, and the user is forced to select one of these intermediate steps as the destination for person-to-person communication instead of the real destination, i.e., the person. If each person had an address, the network could decide the right intermediate device, or the person could dynamically change the device as appropriate in accordance with the requirement.

(vi) Privacy: Privacy and security issues of the current Internet are not concealed to any further extent. So far, we have many serious and proficient attempts and solutions available in the literature, but the truth is that we still receive lot of spam in our mailbox today. Moreover, every day, it is becoming necessary that the next generation of Internet must allow users the option of verification of content, its sources, and its destinations and intermediate systems. In addition, the privacy of data, location, and data integrity should be considered while designing new protocols and communications frameworks.

(vii) New framework integration: As far as the current Internet framework is concerned, data plane, control, and management plans are merging with one another. In today's Internet, the control messages (for example, TCP-connection setup messages) or management messages (such as SNMP messages) follow the same links as data messages. Moreover, control signals are also piggybacked on data packets. This results in high vulnerability and introduces significant security risk as demonstrated by all of the security attacks on the Internet. In contrast, the telephone network uses a separate control network and is generally considered more secure than Internet. For this problem, generalized multiprotocol label switching (GMPLS) is one attempt to separate control and data planes. One benefit of this parting is that it allows data plane to be non packet-oriented such as wavelengths, SONET frames, or even power-transmission lines. This separation is expected to be an integral part of the next-generation Internet architecture.

(viii) Isolation requirements: For many, critical applications—such as military and various monitoring ones—users mostly demand isolation of the application (s) in a shared environment. "Isolation" here means that the performance of one application is not be affected by other application(s) sharing the same resources. Technically speaking, this is difficult to achieve; however, one substitution is to provide dedicated resources to such application(s). To bring this into reality, we have virtual private lines (T1/E1 lines) from telecommunications companies to create private networks. Similarly, it seems that the next-generation networks shall provide users with a programmable mix of isolation and sharing services for application(s).

(ix) Symmetric and asymmetric Internet protocols: Currently, although most Internet protocols are symmetric, they were designed for end-systems with similar competencies. Similarly, regarding wireless sensor and palm-device networks, one can argue that the end-system is significantly resource constrained compared with the others. Thus, in some instances it is reasonable to allow asymmetric protocols, and in the future we expect more sophisticated asymmetric protocols.

(x) Quality of service (QoS): As the name indicates, QoS belongs to a service that in one way or another is related to those packets that affect the services provided. We must know that the ultimate goal of any communication domain is to provide various services to users while using different wireless paradigms. Therefore, the main interest of the user is to receive guarantees about the delay and throughput of data-packet flows. However, IP-based networks make it difficult to guarantee QoS. In contrast, the future-generation Internet should allow a variety of QoS guarantees including total isolation if desired.

References

1. Gromov GR (2002) The roads and crossroads of internet history. NetValley [online]. [cit. 2015-05-22]. Dostupné z: http://www.netvalley.com/cgi-bin/intval/net_historypl (2002)
2. Leiner BM, Cerf VG, Clark DD, Kahn RE, Kleinrock L, Lynch DC, Postel J, Roberts LG, Wolff S (2009) A brief history of the Internet. ACM SIGCOMM Comput Commun Rev 39 (5):22–31
3. Russell AL (2012) Histories of Networking vs. the History of the Internet. In: SIGCIS Workshop
4. Cooper M (2014) The future of internet-enabled innovation: the long history and increasing importance of public-service principles for 21st century public digital communications networks. J on Telecomm & High Tech L 12:1–265
5. Handley M (2006) Why the Internet only just works. BT Technol J 24(3):119–129

6. Stuckmann P, Zimmermann R (2009) European research on future Internet design. Wirel Commun IEEE 16(5):14–22
7. Pan J, Paul S, Jain R (2011) A survey of the research on future internet architectures. Commun Mag, IEEE 49(7):26–36
8. Choi J, Han J, Cho E, Kwon TT, Choi Y (2011) A survey on content-oriented networking for efficient content delivery. Commun Mag, IEEE 49(3):121–127
9. Yu T, Zhang Y, Lin KJ (2007) Efficient algorithms for Web services selection with end-to-end QoS constraints. ACM Transa on Web (TWEB) 1(1):6
10. Floyd S, Fall K (1999) Promoting the use of end-to-end congestion control in the Internet. IEEE/ACM Trans on Netw (TON) 7(4):458–472
11. Fielding R, Gettys J, Mogul J, Frystyk H, Masinter L, Leach P, Berners-Lee T (1999) Hypertext transfer protocol–HTTP/1.1. No. RFC 2616
12. Padmanabhan VN, Mogul JC (1995) Improving HTTP latency. Comput Netw ISDN Syst 28 (1):25–35
13. Stoica I, Adkins D, Zhuang S, Shenker S, Surana S (2002) Internet indirection infrastructure. ACM SIGCOMM Comput Commun Rev 32(4):73–86. ACM
14. Thomasson JK, Neil RT, Davis MM, Mosbarger ML (2015) System and method for multicasting IPSEC protected communications. U.S. Patent 8,953,801. Issued 10 Feb 2015
15. Jokela P, Melen J, Moskowitz R (2015) Using the encapsulating security payload (ESP) transport format with the host identity protocol (HIP)
16. Hakiri A, Berthou P, Gokhale A, Schmidt DC, Thierry G (2014) Supporting SIP-based end-to-end data distribution service QoS in WANs. J Syst Softw 95:100–121
17. Moyer MJ, Rao JR, Rohatgi P (1999) A survey of security issues in multicast communications. Netw IEEE 13(6):12–23
18. Takase A, Tanabe S, Endo N, Takeyari R, Mishina Y, Oouchi T, Yanagi J (1997) Multicast communications method. U.S. Patent 5,612,959. Issued 18 Mar 1997
19. Rodrigues R, Druschel P (2010) Peer-to-peer systems. Commun ACM 53(10):72–82
20. Liben-Nowell D, Balakrishnan H, Karger D (2002) Analysis of the evolution of peer-to-peer systems. In: Proceedings of the twenty-first annual symposium on Principles of distributed computing, pp 233–242. ACM
21. Lua EK, Crowcroft J, Pias M, Sharma R, Sharon L (2005) A survey and comparison of peer-to-peer overlay network schemes. Commun Surv Tutorials, IEEE 7(2):72–93
22. Risson J, Moors T (2006) Survey of research towards robust peer-to-peer networks: search methods. Comput Netw 50(17):3485–3521
23. Detti A, Bruno R, Blefari-Melazzi N (2013) Peer-to-peer live adaptive video streaming for information centric cellular networks. In: 2013 IEEE 24th international symposium on Personal Indoor and Mobile Radio Communications (PIMRC), pp 3583–3588. IEEE
24. Peltotalo J, Jarmo H, Jantunen A, Saukko M, Vaatamoinen L, Curcio I, Bouazizi I, Hannuksela M (2008) Peer-to-peer streaming technology survey. In: Networking, 2008. ICN 2008. Seventh International Conference on, pp 342–350. IEEE
25. Mangili M, Fabio M, Antonio C (2013) A comparative study of content-centric and content-distribution networks: performance and bounds. In: Global Communications Conference (GLOBECOM), 2013 IEEE, pp 1403–1409. IEEE
26. Al-Kanj Lina, Dawy Zaher, Yaacoub Elias (2013) Energy-aware cooperative content distribution over wireless networks: design alternatives and implementation aspects. Commun Surv & Tutorials, IEEE 15(4):1736–1760
27. Xu C, Fallon E, Qiao Y, Zhong L, Muntean G-M (2011) Performance evaluation of multimedia content distribution over multi-homed wireless networks. Broadcast, IEEE Trans on 57(2):204–215
28. Katsaros D (2015) Cache control issues in pub–sub networks and wireless sensor networks. Coordination control of distributed systems. Springer, Berlin, pp 259–264
29. Travers N, Zeinab H, Nelly V, Du Mouza C, Christophides V, Scholl M (2014) RSS feeds behavior analysis, structure and vocabulary. Int J Web Inf Syst 10(3):291–320

30. Ma XingKong, Wang YiJie, Sun WeiDong (2014) Feverfew: a scalable coverage-based hybrid overlay for Internet-scale pub/sub networks. Sci China Inf Sci 57(5):1–14
31. Axelrod RS, VanDeveer SD (eds) (2014) The global environment: institutions, law, and policy. CQ Press
32. Sezer S, Scott-Hayward S, Chouhan P-K, Fraser B, Lake D, Finnegan J, Viljoen N, Miller M, Rao N (2013) Are we ready for SDN? Implementation challenges for software-defined networks. Commun Mag, IEEE 51(7):36–43
33. Brito GM, Velloso PB, Moraes IM (2013) Information-centric networks. Inf-Centric Netw 13–22
34. Ghodsi A, Shenker S, Koponen T, Singla A, Barath R, James W (2011) Information-centric networking: seeing the forest for the trees. In: Proceedings of the 10th ACM Workshop on Hot Topics in Networks, p 1. ACM
35. Loo J, Aiash M (2015) Challenges and solutions for secure information centric networks: a case study of the NetInf architecture. J Net Comput Appl 50:64–72
36. Abidi A, Gammar SM, Kamoun F, Dabbous W, Thierry T (2014) Towards a new internetworking architecture: a new deployment approach for information centric networks. In: Distributed Computing and Networking, pp 519–524. Springer, Berlin, Heidelberg
37. Kim D-H, Kim J-H, Kim Y-S, Yoon H-S, Yeom I (2015) End-to-end mobility support in content centric networks. Int J Commun Syst 28(6):1151–1167
38. Zhang L, Afanasyev A, Burke J, Jacobson V, Crowley P, Papadopoulos C, Wang L, Zhang B (2014) Named data networking. ACM SIGCOMM Comput Commun Rev 44(3):66–73
39. Pentikousis K, Ohlman B, Corujo D, Boggia G, Tyson G, Davies E, Molinaro A, Eum S (2015) Information-centric Networking: Baseline Scenarios. No. RFC 7476
40. Xu Y, Li Y, Lin T, Wang Z, Zhang G, Tang H, Ci S (2014) An adaptive per-application storage management scheme based on manifold learning in information centric networks. Future Gener Comput Syst 36:170–179

Chapter 2
Information-Centric Networks (ICN)

**Muhammad Azfar Yaqub, Syed Hassan Ahmed,
Safdar Hussain Bouk and Dongkyun Kim**

Abstract During the past decades, serious efforts have been made to propose various architectures for the future Internet. Each of those architectures has one thing in common, i.e., to focus on content delivery rather than on host-centric approaches. However, only few of them gained popularity due to their possible applications being investigated. In this chapter, we describe the overview of various future Internet architectures such as data-oriented networking architecture, content-centric networking, named-data networking, publish/subscribe, and network of information. The main objective of this chapter is to allow our readers to become familiar with the transformation of these architectures.

Keywords ICN · DONA · CCN · NDN · Pub/sub · Net-Inf

2.1 Information-Centric Network (ICN)

2.1.1 Brief History

The core idea behind information-centric networking (ICN) architectures is that who is communicating is less significant than what data are required. This paradigm shift has occurred due to end-users' use of today's Internet, which is more content-centric than location-centric, e.g., file sharing, social networking, or retrieval of aggregated data. The ICN concept was initially proposed in TRIAD [1], which proposed name-based information communication. Since then, researchers have proposed multiple architectures (Fig. 2.1). In 2006, the data-oriented network architecture (DONA) project [2] at UC Berkeley proposed an ICN architecture, which improved the security and architecture of TRIAD. The Publish Subscribe Internet Technology (PURSUIT) [3] project, a continuation of the Publish Subscribe Internet Routing Paradigm (PSIRP) [4] project, both funded by the EU Framework 7 Program (FP7), have proposed a publish/subscribe protocol stack that replaces the IP protocol stack. In another approach, the Network of Information (NetInf) project [5] was initially proposed by the European FP7 4WARD [6]

S.H. Ahmed et al., *Content-Centric Networks*, SpringerBriefs in Electrical and Computer Engineering, DOI 10.1007/978-981-10-0066-9_2

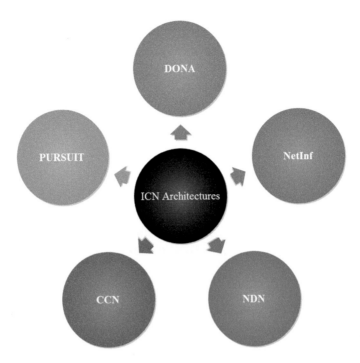

Fig. 2.1 ICN architectures

project, and further development has been made by the Scalable and Adaptive Internet Solutions (SAIL) [7] project. Similarly, Van Jacobson, a Research Fellow at PARC, proposed the Content Centric Networking (CCN) project [8] in 2007. Currently work is being performed to enhance the CCN architecture called "named-data networks" (NDN) [9].

All of these approaches differ in terms of implementation, but they have the same goal, i.e., to improve the performance and end-user experience of the Internet by providing access to content and services by name rather than by original location. This is achieved by changing the concept of link protection to content protection and by exploiting in-network storage of content. In addition, traditional networks also benefit from ICN technologies, i.e., content delivery networks, which aim to distribute content efficiently and swiftly.

2.1.2 ICN Core Architectures

In this section, we give an introduction and overview of four ICN approaches in networking, namely, CCN/NDN, DONA, NetInf, and PURSUIT. A higher level description of these architectures is illustrated to provide a general understanding of the readers

2.1.2.1   CCN/NDN

Introduction

The content-centric network (CCN) architecture was originally proposed by Van Jacobson as a project initiated by Palo Alto Research Center (PARC). Before publishing [8] the architecture, Jacobson first introduced it in his Google Tech Talk [10]. The main idea behind CCN is that the request (Interest) packets containing the desired content/data name are broadcast by a consumer node, and routing protocols are employed to distribute information about the location of content based on the name using the longest prefix matching. Routing aggregation is leveraged through a hierarchical naming scheme. The content provider, or any other network node with a copy of the requested content, routes the required content, along with additional authentication and data-integrity information, along the interests reverse route. Furthermore, caching on each path node is enabled depending on the caching policy of the node. An overview of the communication is illustrated in Fig. 2.4. In an intermittent-connectivity scenario, this can speed up content retrieval because the content is replicated in the network [8]. Further details—such as naming, security, caching, name resolution and routing, transport, and mobility of the CCN architecture—are discussed in detail in the next chapter.

Named-data networking (NDN) is an enhanced version of the CCN architecture. Similar to CCN, NDN also follows the interest/data packet combination to obtain any particular data. There are, however, some architecture differences incorporated into NDN, which reduces the interest/data search time as well as the interest looping issue. Figures 2.2 and 2.3 illustrate the difference in both CCN and NDN basic operations (Fig. 2.4).

Naming

NDN adopts a hierarchical naming scheme, e.g., information may have the name/work/class101/presentation.pdf, where the sign "/" shows the hierarchy of the name component. The relationships and context of the data elements are easily represented in this hierarchical structure. In a typical CCN, each node consists of three data structures: a pending interest table (PIT), a content store (CS), and a forwarding information base (FIB). PIT contains a list of pending and satisfied interests. The entries include a content name, the interest-incoming interface, a NONCE value to identify the individual interest packet, and timers for PIT-entry management. CS provides a cache to store the content available at the node and content received from other nodes based on the caching policy of the node. FIB helps in routing the incoming interest to the next hop toward the content provider; it maintains name prefixes and outgoing interfaces for interest packets. In addition, a forwarding algorithm is used to provide a forwarding strategy, which uses these data structures.

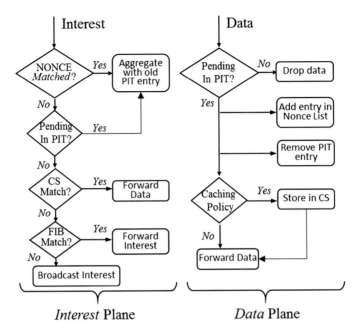

Fig. 2.2 CCN

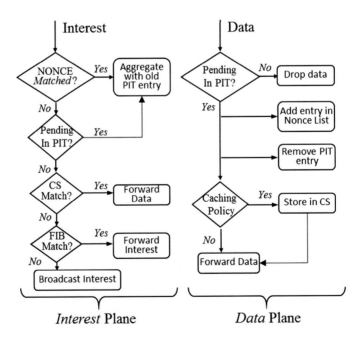

Fig. 2.3 NDN

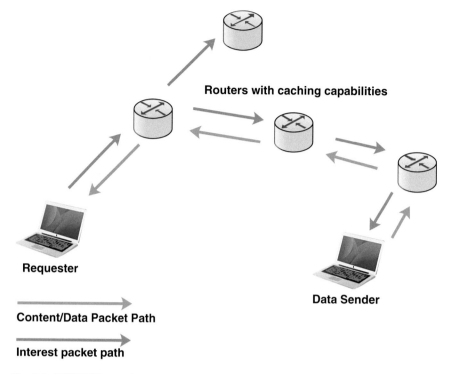

Routers with caching capabilities

Requester

Data Sender

Content/Data Packet Path

Interest packet path

Fig. 2.4 CCN/NDN overview

Security

In NDN, the content publisher provides security by cryptographically signing each data packet [8]. Hierarchical namespace is used to achieve better routing scalability. To achieve data integrity, every content is signed with the publisher's secret key, but the trust in the signing key must be established through some external means. The naming in CCN/NDN typically does not contain the publisher's key (PK). Although this helps with the human readability of the names, self-certification is not possible. Multiple methods are used to verify the key, such as information through a friend, direct information, information through a trusted third party, or information through a global PKI.

Routing and Name resolution

In CCN/NDN, name-based routing is used to forward packets between the source and destination. The client/requester broadcasts interest packets for the required content in the network. This interest is forwarded to the name-prefix of the destination using longest-prefix matching at the FIB of each intermediate node.

Each incoming interest's information is stored in the PIT; furthermore, multiple requests of the same content are aggregated together. When the requested content copy is found at a node, a data packet is sent back with the requested data on the reverse path toward the client. In addition, each node along the data path can cache a copy of the data.

Caching

Caching is one of the major advantages of the ICN architectures [11]. Caching the content in the CS of a node is analogous to the buffer memory in IP routers; however, the IP routers cannot reuse the data packet after forwarding. However, in NDN the storage of packets is possible at each NDN node, thus allowing the node to satisfy any future request for the particular data. In addition, as the content name does not contain any information of the user, thus making the users more secure.

Transport

The CCN/NDN architecture does not provide any transport-layer functionality. The transport-layer functionalities are provided by the application or some supporting libraries as well as the forwarding algorithm used in NDN architecture. The hierarchical namespace allows the information required for transport to be included in the content name, thus eliminating the need for transport-layer information such as sequence and port numbers. The application itself monitors the state of each outstanding interest in the PIT. After a certain timeout, retransmission is initiated. To limit congestion in the network, each interest packet has a limited lifetime; furthermore, caching the data packets at each node mitigates any congestion losses in the network because retransmitted interest packet will be satisfied by the node with the particular data packet in its cache.

2.1.2.2 DONA

Introduction

DONA's architecture involves a redesigning of the current Internet naming, i.e., DNS names are replaced with flat, self-certifying names, and DNS name resolution is replaced with any cast name resolution process. Furthermore, these changes are incorporated above the IP layer, thus leveraging the lower layers of path discovery mechanisms. The architecture provides improved data retrieval as well as improved service by providing persistence, authentication [12, 13], and availability.

In DONA, the source/content provider is responsible for publishing the content in the network. To serve data, the nodes must authorize with the resolution infrastructure. A route-by-name paradigm is used for name resolution. Now, instead

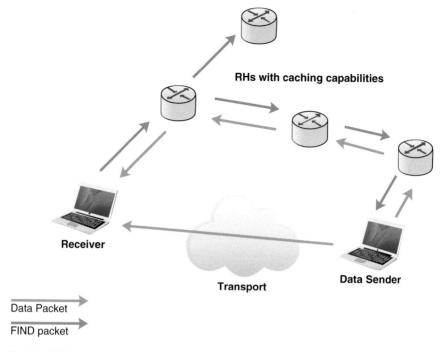

RHs with caching capabilities

Receiver

Transport

Data Sender

Data Packet

FIND packet

Fig. 2.5 DONA overview

of using DNS servers, DONA relies on the network entities called "resolution handlers" (RHs). The request (FIND) packets are forwarded through multiple RHs toward the node with a copy of the content as illustrated in Fig. 2.5. The content/data can be acquired through two methods: (1) it is sent back through the same path the interest packet came in on with caching enabled on each encountered RH or (2) it can be sent back directly toward the consumer. The source also has the option to register their principals with the RH so that the request packets can be sent to them directly. However, the registrations must be renewed periodically. RH routes requests using a hierarchical approach to find the closest content provider. The any cast name resolution used in DONA provides support for the network middle boxes (e.g., firewalls, net address translators, proxies, etc.) by providing a separate mechanism for path discovery.

Naming

In DONA, a flat namespace is used, and the names are organized by using principals. A public private key pair is used to associate each principal, which is in the form P:L where P and L are the globally unique principal fields containing the cryptographic hash of the publisher's public key and the object label, respectively.

Each datum received contains metadata with a minimum of three fields, i.e., data, public key, and signature. To ensure data integrity, the requesting client relies on the principals' signatures. In addition, because P is unique for each publisher, republishing the same content by a different publisher will result in multiple copies of the same content. These multiple copies can either be removed by using various methods, i.e., wildcard queries or principal delegation or be used to satisfy multiple content requests [14].

Security

The self-certifying namespace used in DONA attributes to name-data integrity. This removes the necessity for PKIs. This is achieved by adding the cryptographic hash function of the content in the object label "L." In the event of dynamic data, the signature of the contents hash is added in the metadata of the content; furthermore, the public key corresponding to the hash in the ID's authenticator field is used to sign the metadata. This allows the object label to be securely bound with the data and also provides information to handle the data yet to arrive/exist. In contrast, using self-certifying names provides a trade-off between the human readability of the content names and name-data integrity. However, self-certifying names eliminates the need for a PKI by allowing the comparison of the receiving data identity with the one sent in the data request, consecutively making the security process simple and reliable (offline data verification).

DONA architecture relies on the IP-level mechanisms and contractual limits set by the providers to restrain the unwanted packets from overwhelming the server, the client, or the RH.

Routing and Name Resolution

As discussed previously, a name-based routing paradigm is used for name resolution in DONA. This is achieved through the network entities called "resolution handlers" (RHs) instead of DNA servers. There is at least one RH available in each domain. Each node in the domain has information on the RH through local configuration information. The client and RH use two primitives, FIND (P:L) and REGISTER (P:L), to achieve name resolution. The nodes willing to serve data use the REGISTER (P:L) packet to register the datum with the RH. Each RH maintains a registration table, which is used to map the incoming requests to the destination of the content, copy, or to the next hop RH toward the content copy. The RHs are organized in a hierarchical manner.

In the event that the RH receives a new REGISTER (P:L) packet from a child node, it is stored in the registration table, and the RH also forwards the packet to its parent and peers. The peer receiving the REGISTER will forward the packet on the basis of the local policy. The REGISTER packet is not forwarded onward if a record already exists or if the new REGISTER comes from a copy further away from the previous copy.

The client issues a FIND (P:L) packet to locate the content named "P:L." If there is an entry in the registration table when the request is received at the RH, the request is forwarded to the next hop RH; otherwise the RH forwards the request to its parent RH. In case of more than one entry being in the registration table, the closest one is selected. Once a content copy is found, the data are either routed back to the client by way of the reverse request path with caching enabled on the RHs or forwarded directly to the client as shown in Fig. 2.5. In DONA, name matching is accomplished using longest-prefix matching [15].

Caching

In DONA, RHs can be enabled with a universal-caching mechanism. To enable caching, the RH populates its cache by replacing the source IP address and port number of the FIND packet with its own and then forwarding it to the next-hop RH, thus, thus ensuring the traversal of the response packet through this RH. The RH stores the data in its cache before forwarding it to the requesting node. All the data items in the cache are labeled with a TTL or some other validation metadata, which ensures the time period of the data.

When a FIND arrives and there is a cache match, the RH will initiate the appropriate transport response, which will lead to the standard application-level exchange of data. In case the transport or application-level protocol information in the FIND is ambiguous to the RH, then it does not provide any caching for that particular request.

Transport

The DONA architecture relies on the existing transport protocols, i.e., TCP, to provide the forwarding mechanisms and other transport functionalities such as flow control, congestion control, and reliability.

2.1.2.3   NetInf

Introduction

Like the DONA architecture, NetInf also uses flat namespace [16]; hence, a public key infrastructure (PKI) is not required. NetInf's content model is based on the widely used multipurpose internet mail extensions (MIME) standard. Furthermore, search primitives, which provide links between the search item and the object name, are also a part of the architecture. Two objects-retrieval approaches are offered in the architecture, namely, name resolution and name-based routing. Depending on

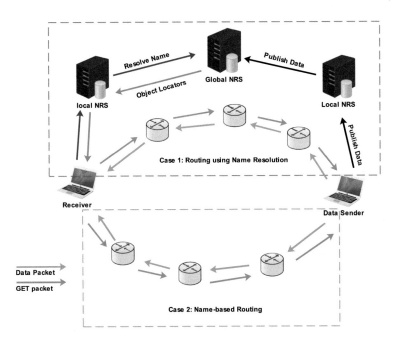

Fig. 2.6 NetInf overview

the model used, the source/NetInf node can either register with the name resolution service (NRS) to publish the content (termed "named-data objects" [NDOs]) or use a routing protocol to announce the routing information.

As illustrated in Fig. 2.6, in case no. 1, the client first forwards the request to the NRS, which gives the available locaters of the particular NDO name; subsequently, the client retrieves a copy of the data from the best available sources. Alternatively, using name-based routing, the client can directly send out an NDO GET request, which is forwarded to the source. The data are sent to the client as soon as a copy of the NDO is reached. These can either be used separately in the network or merged into a hybrid scheme, in which case switching between the two schemes is performed on hop-by-hop basis. This hybrid scheme allows NetInf to adapt and scale itself to the different requirements in the network such as network mobility [17], delay-tolerant networking (DTN), and global connectivity. Furthermore, NetInf architecture can be deployed as an extra layer on top of the existing network infrastructure, thus simplifying the migration of applications to the new infrastructure.

Naming

NetInf uses a flat namespace with a structure similar to the DONA namespace. NetInf aims to accommodate different ICN architectures by deploying the naming such that the naming aims to differentiate three aspects: (1) the common naming format used by each node; (2) security related information to maintain the integrity of the data; and (3) name-object binding validation mechanisms. The NetInf naming format supports different hashing schemes. The owners public key hash digests is also contained in the name to support the data that is yet to arrive. Furthermore, different naming representation is supported, i.e., Uniform Resource Identifier (URI) and binary representation.

Security

NetInf uses the self-certifying namespace, which provides object security service for static as well as dynamic objects. The aforementioned naming format and the object model enable data-integrity validation by the nodes. Like DONA, NetInf validation of the named data can be achieved without the PKI infrastructure. In addition, object security is provided through public key cryptography, the pseudonym of the owner, and identification.

Name resolution

To incorporate the different ICN architectures, NetInf supports a variety of name-resolution services. NetInf merges name resolution and name-based routing to retrieve the data. A new interdomain interface is defined for name resolution and routing, which allows different schemes to be applied in multiple parts of the network. Today's URLs are supported by NetInf name resolution; hence, NetInf can be integrated smoothly with the current infrastructure. Multiple types of NRS are supported in NetInf such as both local and global NRS. Using local NRS, the operators can reduce and control the traffic flow thus potentially decreasing the load on caches and servers.

A number of name-resolution mechanisms have been developed such as multilevel distributed hash table (MDHT) [18], hierarchical skipnet (HSkip) system [19] and late-locator construction (LLC) [20]. MDHT and HSkip systems provide a global and hierarchical NRS that is topologically embedded in the core network to improve stability, scalability, copy-locator selection, and efficient data dissemination. LLC focuses on high-dynamic network topology handling, which includes movable networks. This NRS approach allows a smooth transition from the Internet while using the current infrastructure. For example, traditional URLs can be resolved from object names and retrieved using the existing HTTP protocol.

Caching

Caching plays an important role in efficient content distribution in NetInf. NetInf supports three caching options: on-path caching, off-path caching, and peer caching.

The NetInf router has a built-in caching feature to enable on-path cache, which caches objects while routing objects in response to the GET request. Off-path cache is placed in the network to reduce the traffic and latency. This cache is not directly in the request/data path. It is typically connected to an NRS in the network. The cache broadcasts the cached objects to the NRS and, based on the popularity, the NRS informs which data to cache. In this way, the off-path cache can avoid the steps to obtain the information from GET requests of the requested objects. In peer-caching, the NetInf nodes can function as an on/off-path cache. The peers can broadcast the cached data in the network. The NRS can route the GET request, thus reducing interdomain traffic and latency and additionally minimizing the load on the data servers.

Transport

In NetInf, different forwarding mechanisms are used to retrieve a data object, locator, or redirection hints using request/response messages. This communication is managed by convergence layers (CLs), which provide a concrete abstraction of the NetInf protocol from the lower layers. Thus, by ensuring smooth NetInf implementation across technologies. CL ensures that communication between the NetInf nodes is achieved in a hop-by-hop manner. An example of the CL system is given in [21]. The CL implements a specific transport protocol, which manages the resources required for sharing and reliability of corresponding network paths.

2.1.2.4   PURSUIT

Introduction/Model

The Publish-Subscribe Internet Technology (PURSUIT) project was previously known as the Publish-Subscribe Internet Routing Paradigm (PSIRP) [22]. In PURSUIT, sources publish the contents into the network as shown in Fig. 2.7. The receivers can subscribe to the published contents through the rendezvous systems. A rendezvous system helps in locating the scope and publications in the network. Each piece of the published content belongs to a specific named scope. The subscription requests contain the scope identifier (SI) and the rendezvous identifier (RI), which together identify/name the particular desired content. Using these identifiers in a matching procedure results in a forwarding identifier (FI), which is used by the source to forward the data. A bloom filter [23] is specified in the FI, which is used by the intermediate routers to select the interfaces to forward contents as shown in Fig. 2.7. This relieves the router from maintaining the forwarding states. However, a bloom filter yields some false-positive results, thus leading to forwarding on interfaces where there are no receivers.

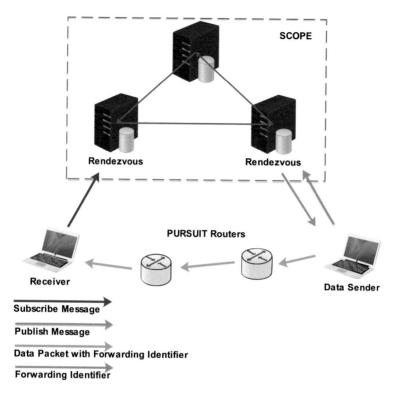

Fig. 2.7 PURSUIT overview

Naming

PURSUIT uses a flat namespace with two types of names, namely, RI and SI. These identifiers together establish the name of the content. RIs help in mapping the content between publishers and subscribers. In addition, the forwarding identifier (FI) is used by routers to identify the path from the publisher to the subscribers.

Security

Security is an integral aspect of PURSUIT design employed to avoid inherent security glitches in the network. PURSUIT uses self-certifying names, which alleviate the need for a PKI; therefore, nodes can easily check the name-data integrity based on the received data's name. The other security aims are to avoid unwanted traffic on both the rendezvous and forwarding layers; furthermore, policy should be enforced such that only valid subscribers can obtain the data. PURSUIT makes use of elliptic-curve cryptography (ECC) [24] for signature verification and packet-level authentication (PLA) [25] to provide network layer confidentiality, authenticity, and accountability of the data.

Routing and Name Resolution

The routing management is responsible for selecting the best interdomain-route forwarding of the publications. PURSUIT uses the resolution model called a "rendezvous point." The name resolution of the data is performed at this point. However, the data return path to the subscriber does not have to include the rendezvous point. Forwarding is performed using the source routing approach using bloom filters called "zfilters" carried in the FI. The bloom filter describes the route from the source to the destination because it contains all the names of the routing links. This information is attached to the data as the FI. At each node, the router checks whether or not the link identifier is present in the packet using a simple AND operation. Thus, in PURSUIT, increasing packet length as well as the network resources are reduced.

Cache

Caching in PURSUIT is mainly provided as a dedicated solution to a problem for which caching might offer some benefit. Moreover, multiple caches of an object can be maintained based on the scope of the rendezvous point for the identifier associated with the object.

Transport

PURSUIT's basic forwarding process is based on the bloom filters as mentioned previously. Each object has a unique algorithmically derived name from the original name, which helps to handle the flow control. Alternatively, subscribers can add flow-control feedback in an algorithmically derived name to which the source can subscribe.

   In addition to the other initiatives taken by the research community for future Internet architectures, the CCN got much attention recently. Therefore, we discuss CCN in detail in the following chapters and will also provide research challenges for this subject.

References

1. Cheriton D, Gritter (2000) Triad: a new next-generation internet architecture
2. Koponen T, Chawla M, Chun B-G, Ermolinskiy A, Kim KH, Shenker S, Stoica I (2007) A data-oriented (and beyond) network architecture. SIGCOMM Comput Commun Rev 37 (4):181–192
3. FP7 PURSUIT project (Online). Available: http://www.fp7-pursuit.eu/PursuitWeb/
4. FP7 PSIRP project (Online). Available: http://www.psirp.org/

5. Dannewitz C, Kutscher D, Ohlman B, Farrell S, Ahlgren B, Karl H (2013) Network of information (netinf)–an information-centric networking architecture. Comput Commun 36 (7):721–735
6. FP7 4WARD project (Online). Available: http://www.4ward-project.eu/
7. FP7 SAIL project. [Online]. Available: http://www.sail-project.eu/
8. Jacobson V, Smetters DK, Thornton JD, Plass MF, Briggs NH, Braynard RL (2009) Networking named content. In: Proceedings of the 5th International Conference on Emerging Networking Experiments and Technologies, CoNEXT '09, pp 1–12, New York, NY, USA, ACM
9. Zhang L, Afanasyev A, Burke J, Jacobson V, Claffy K, Crowley P, Papadopoulos C, Wang L, Zhang B (2014) Named data networking. SIGCOMM Comput Commun Rev 44(3):66–73
10. Jacobson V (2006) A new way to look at networking, Google Tech Talk, Aug 2006
11. Fayazbakhsh SK, Lin Y, Tootoonchian A, Ghodsi A, Koponen T, Maggs B, Ng K, Sekar V, Shenker S (2013). Less pain, most of the gain: incrementally deployable ICN. SIGCOMM Comput Commun Rev 43(4)
12. Mazi`eres D, Kaminsky M, Kaashoek M.F, Witchel E (1999) Separating key management from file system security. In: Proceedings of SOSP '99, pp 124–139, Charleston, SC, USA, Dec 1999
13. Moskowitz R, Nikander P (2006) Host Identity protocol architecture. RFC 4423, IETF, May 2006
14. Xylomenos G, Ververidis CN, Siris VA, Fotiou N, Tsilopoulos C, Vasilakos X, Katsaros KV, Polyzos GC (2014) A survey of information-centric networking research. Commun Surv Tutorials IEEE 16(2):1024–1049 (Second Quarter 2014)
15. Ghodsi A et al (2011) Naming in content-oriented architectures, In: Proceedings of ACM SIGCOMM workshop information-centric networking, Toronto, Canada, Aug 2011
16. Dannewitz C et al (2010) Secure naming for a network of information. In: Proceedings of 13th IEEE global internet symposium '10, San Diego, CA, Mar 2010
17. Eriksson A, Ohlman B (2007) Dynamic internetworking based on late locator construction. In: 10th IEEE global internet symposium, 2007
18. D'Ambrosio M, Dannewitz C, Karl H, Vercellone V (2011) MDHT: a hierarchical name resolution service for information-centric networks. In: Proceedings of ACM SIGCOMM workshop on information-centric networking, ACM, New York, NY, USA, 2011, pp. 7–12
19. Dannewitz C, D'Ambrosio M, Karl H, Vercellone V (2013) Hierarchical DHT-based name resolution for information-centric networks, Elsevier computer communications, special issue on information-centric networking, 2013
20. Eriksson A, Ohlman B (2007) Dynamic internetworking based on late locator construction. In: 10th IEEE global internet symposium, 2007
21. Kutscher D, Ahlgren B, D'Ambrosio M, Davies E, Eriksson AE, Farrell S, Grönvall B, Imbrenda K, Kauffmann B, Kunzmann G, Lindgren A, Marsh I, Muscariello L, Ohlman B, Persson K-A, Pöyhönen P, Shehada M, Staehle D, Strandberg O, Tuononen J, Vercellone V (2012) (D.3.2) Content delivery and operations, deliverable, SAIL 7th FP EU-funded project, May 2012
22. Ain M et al (2009) D2.3–Architecture definition, component descriptions, and requirements, deliverable, PSIRP 7th FP EU-funded project, Feb 2009
23. Bloom BH (1970) Space/time Trade-offs in hash coding with allowable errors. ACM Commun 13(7):422–426
24. Miller VS (1985) Use of elliptic curves in cryptography. In: Proceedings of CRYPTO '85: the advances in cryptology. Aug 1985
25. Lagutin D (2008) Redesigning internet-the packet level authentication architecture, Licentiate's thesis, Helsinki University of Technology, Finland

Chapter 3
Content-Centric Networks (CCN)

Syed Hassan Ahmed, Safdar Hussain Bouk and Dongkyun Kim

Abstract Several initiatives have been taken in the past decade to improve the architecture and performance of information-centric networks (ICN). In this context, Van Jacobson presented a new architecture called the "content-centric network" (CCN), which was funded by the PARC research company. The main goal of CCN was to change host-centric communication into content-centric communication. In CCN, the requester is known as a "consumer" who sends an "interest" to the network, and any node with the requested data can send back the "content" to the consumer by way of the same path. This simple overview seems superficial without explanation. Therefore, in this chapter, we provide readers with the history of CCN followed by its basic operations. Moreover, we describe the different components of CCN in detail such as the following: (1) What constitutes "content"? (2) What is the structure of content and an interest message? (3) How can the interest can be forwarded and, in response, how can data retrieval be efficient compared with the current Internet architecture? We believe that this chapter will enable our respective readers have a solid background about the CCN and, in later stages, that they can become active researchers in the given field.

Keywords CCN · CCNx · Content · Data · Interest · Content retrieval · Applications

3.1 Introduction

The Content Centric Networking (CCN) [1] Project was started and managed by the PARC (Palo Alto Research Center) with the aims to develop a flexible, simple, and universal next-generation communication architecture. It is also targeted that CCN will alleviate communication complexities, will require less configuration, must be efficient and scalable, and will have the application design patterns of the current communication technologies. The implementation of CCN is called "CCNx," and the project has gone through different stages, such as the preliminary implementation of CCNx and currently different issues are being researched. The next phase of this project is the formalization of different standardization proposals that plan to

© The Author(s) 2016
S.H. Ahmed et al., *Content-Centric Networks*, SpringerBriefs in Electrical and Computer Engineering, DOI 10.1007/978-981-10-0066-9_3

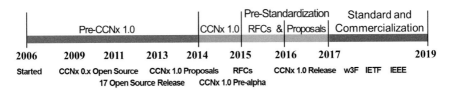

Fig. 3.1 CCN timeline

standardize CCN in 2017 to 2018. The detailed timeline of the CNN Project is shown in Fig. 3.1.

It is stated by the Jacobson [2], who coined the concept of CCN, *"that any architecture designed to run over anything is necessarily an overlay. The things that matter are the capabilities."* The argument was based on the historical overlay technologies, such as IP technology, which was initiated as an overlay on the phone system; in turn, currently the phone system is used as an overlay on IP technology. IP in its entirety is independent of any layer in the current Internet technology. Likewise, CCN demonstrates and envisions the same characteristics as an overlay that can run over any technology, including IP, and the converse is also possible, i.e., anything can run over CCN including IP. A similar project was started and funded by the NSF (National Science Foundation) in mid-2010, under the Future Internet Architecture program, called "named-data networking" (NDN) [3]. NDN follows the basic CCN functionality with some modifications in the forwarding daemon implementation.

Basically, CCN is one of the future Internet architectures to provide location-independent communication of data, contents, information elements, streaming contents (video/audio), or content segments using content *names* [note that the terms "data," "contents," "content objects," or "information elements" are interchangeable in the context of this book; similarly, a complete content or content object can be a combination of multiple fragments]. It is considered that the content name must uniquely identify the complete content or any fragment(s) of the content and that it is independent of the location information. The content communication is also performed using these names. Thus, it is the center of the CCN architecture. The implementation of CCN [1, 4], i.e., CCNx, provides multiple services including end-to-end, loop-free, multi-hop, multi-path communication, flow control, palpable and impulsive multicast service, security intrinsic to content rather than the connection, content integrity irrespective of the delivery path, distributed caching in the network, etc. The terms "content" or "data" are referred as "ContentObjects" in CCNx. CCN relinquishes the execution of application notwithstanding the type and nature of the lower-layer technology.

End-to-end communication is provided by CCN applications using CCN as an innate part of the application to avoid a more layered complexity. A simplified pull-based or receiver-initiated content communication is adapted in basic CCN architecture. However, a few proposals have adopted push-based CCN architecture [5] suitable for specific applications, e.g., Internet of Things (IoT) applications [6],

Fig. 3.2 CCN daemon

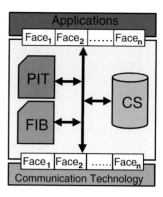

wireless-sensor network (WSN) applications [7], etc. In a pull-based communication, the receiver node or device sends a content request message (called an "interest message" in CCN terminology), and the node with requested data replies with the data message. In CCN, the requesting node is called the "consumer node," and the request satisfying or data bearer node is called the "provider node." All CCN-enabled nodes implement the forwarding and content-caching capabilities (Fig. 3.2).

CCN uses two elementary messages to achieve the simplified pull-based communication, i.e., "*interests*" and "*data*" messages [8]. Along with these two messages, some data structures are maintained at each node to properly forward interest data messages in the network including pending interest table (PIT), forwarding information base (FIB), and the content store (CS). The consumer node sends the interest message if it requires any content, and the provider node replies with the data message with the content or a fragment of the content. The interest message contains the content name and other information to properly *identify* (not *locate*) the content. All of the intermediate nodes that receive the interest message first search in their CS, and if they do not find the specified content, they make an entry in the PIT. Every PIT entry contains content name, incoming face, and other information of every received interest. Then, the intermediate node searches the content name within the FIB to find the outgoing face. The outgoing face linked to the entry with the longest prefix match is used to forward the interest message. When an intermediate node receives the data message, it may hold the copy of the content message depending on whether a caching policy is implemented. Following is the brief description of the data structures and concepts partly mentioned previously that are used by the CCN node to achieve loop-free forwarding and caching in the network.

- **Face**: The term "face" is taken from the word "interface." The face in CCN is nothing but a simplified concept of communication or application interface. This means that the information is exchanged through face(s) between the CCN core (CCN forwarding daemon as well as the data structures used by the CCN) and the software or application(s) running on the node and/or the communication interface(s) installed on the CCN-enabled node. In the latter case, the CCN node configures the communication interface(s) in such a way so that the node can receive and send messages to and from the other nodes in the network. Communication in CCN is mostly either broadcast or multicast in nature; however, there are the cases where point-to-point communication is performed using these faces. The application faces are used by the CCN to communicate with the application instance(s) or process(es) running on the node.

- **Pending interest table (PIT)**: This is a data structure that stores information about unsatisfied interests including content name, incoming face from where the interest message was received, and timer(s). The term "unsatisfied" means that the node has forwarded the interest message received from the downstream and forwarded it to the upstream; however, the interest forwarder is awaiting the response. The timers are used to limit the duration of PIT entry and, ultimately, the size of the PIT. The entries in PIT are organized in such a way that insert, search, delete, and modify functions can be performed easily. An entry is created in the PIT when a node receives an interest message that it cannot satisfy with the requested content. This means that the content is not available in the CS. The pending interest recorded is purged from the PIT in two cases: (1) the node received the content or data in response to that interest message or (2) the message time has expired. Before the creation and deletion of the PIT entry, a lookup operation is performed to make sure that the entry is still present or not. This lookup mechanism uses the longest prefix matching algorithm on *names* to find the entry with the best-matching name in the PIT. Therefore, entries in the PIT should be managed in such a way that it should expedite the lookup process because there may be a huge number of requests in the network. In technical documents, it is also suggested that the PIT should also keep track of the interest's scope (the maximum hop limit, i.e., a two-byte field in the interest message). If an interest with a shorter hop limit is stored in the PIT and a similar interest with a larger hop limit arrives, then the forwarder node must forward the shorter-hop interest and update the PIT entry accordingly. A large random number, called "NONCE," is used in the interest message to uniquely identify it. The NONCE value is stored in the PIT to detect the interest loop as well. If a node receives an interest with similar-value NONCE, it is dropped, and no further action is required. A detailed illustration of a PIT is shown in Fig. 3.3.
- **Forwarding information base (FIB)**: The FIB is analogous to the IP table maintained on Internet routers. When an interest is unsatisfied by the node (i.e., there is no matching content in the CS and no entry in the PIT), then it is forwarded upstream toward potential providers using FIB. Every entry in the FIB is the tuple of name prefix and outgoing face(s). A single prefix entry may have more than one outgoing face associated with it. The interested content name is searched within the FIB using longest-prefix matching. The interest is forwarded to the face associated with resultant entry of the LPM search. A simplified FIB is shown in Fig. 3.4.
- **Content store (CS)**: This is a buffer memory or cache used to store full or partial contents or data packets; however, it is not a persistent storage. The

Fig. 3.3 Details of pending interest table

Name Prefix	Incoming Face ID
`ccn://knu.ac.kr/monet/members/bouk`	2

Parameters				
Nonce	Last Refreshed	Life-Time	Scope	....
`0xA1B3...`	1	2	5	....

Fig. 3.4 FIB structure

Prefix	Outgoing Face ID
	.....
`ccn://knu.ac.kr/monet/`	1
`ccn://comsats.eud.pk/ee/`	1,2
	.....

contents are stored in the CS per cache policies to satisfy future interest messages. Two types of policies for the CS have been discussed: replacement policies and *caching-decision* policies. These policies are important to maximize caching and content-satisfaction efficiency. Cache-replacement policies effectively use the CS capacity. It indicates that when a new content is received, the CS should replace it with a certain content: the content that is least recently used (LRU) or least frequently used (LFU), random content, content with the maximum time in the CS (i.e., FIFO [first in first out]), etc. In contrast, caching-decision policies decide whether or not the received content should be cached. Caching-decision policies may include leave copy everywhere (LCE), leave copy down (LCD), fixed probability p caching, most popular caching (MPC), leave copy on the edge, etc. [9–14]. The content store also uses a 1-bit "stale" flag with each content object. A high stale flag indicates that content will not be sent in response to the interest message and that the interest message should be forwarded to upstream for other potential providers. This bit is set high when the content's freshness time has expired, and the content will be given priority to be eliminated first when the cache reaches to its capacity.

The CS is organized in such a way as to efficiently retrieve contents based on the prefix lookup using content names and the selectors. That is the reason the content table is built on a **name_tree** structure that is keyed by the flat-name representation of the content name.

The same structure can also be used as a basic structure that requires or depends on prefix-matching operations [15], i.e., FIB entries, PIT entries, statistics used by the strategy layer, name enumeration, and creation/deletion notifications. Most of the CCN data structure requires a longest-prefix lookup, and the name_tree structure is inherent in several data structures. Readers interested in a

Fig. 3.5 CS structure

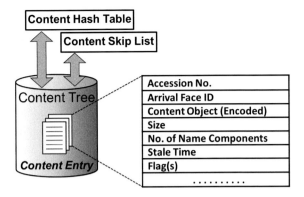

detailed description of name_tree implementation are advised to refer the name_tree structure in the CCNx. The structure of the content store is shown in Fig. 3.5.

A "hash table" is maintained to store content information, and a separate "skip list" is used to orderly maintain the content objects and is used over the stored content objects. The content hash table stores all the content entries. Its "key" type is a portion of the data or the content message other than the actual content object.

"Content entry" represents a single entry in the content store and contains the following:

- The accession number (accession) is a unique identifier assigned to each unique content stored in the content store. Accession numbers are assigned in the order of arrival, and the highest number is assigned to the content that has been received most recently. The most recent number is saved in the CCNx data structure to keep track.
- Face no. or ID from which the content object has arrived initially.
- Size of the content object.
- Index and number of the name components (comps, ncomps)
- Time in seconds before the content object becomes stale.
- Flags, e.g., whether or not the content object is stale.

Name Skip List

The skip list uses the content name to index content entries in an increasing order. The maximum depth of this skip list is up to 30 [readers are advised to refer to the information regarding skip list from the plethora of online and offline resources because skip lists are out of the scope of this book]. The traversal of the skip links using pointers to the next entries will walk through the names in increasing order.

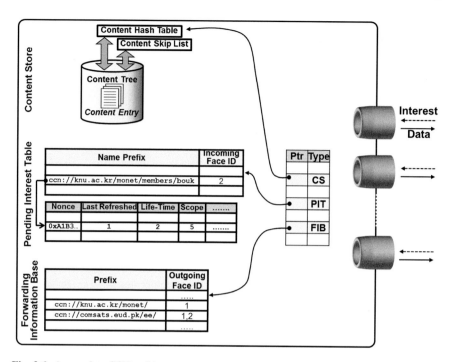

Fig. 3.6 A complete CCN architecture

Accession Skip List

The accession skips list indexes content entries using the accession number information. The content entries may use an accession number, such as (Accession Base + Accession window), to access the content entry.

There may be a case when a content or data message is received by a node that has no previous PIT information or the node did not receive the interest message requesting that content message. This type of content object is called "unsolicited data," which may also be stored in a separate list. Accession numbers of the unsolicited content objects are separately recorded in the data structure.

In CCNx, all three structures, i.e., FIB, PIT, and CS, are connected through a single index, which may result in alleviation of the lookup cost of each message received, processed, and forwarded by the CCNx. This front index is optimized and ordered in such a way as to efficiently perform the operations specific to the received message type as shown in Fig. 3.6.

3.2 Basics of CCN

In CCN, a consumer node sends an interest message if it requires a content. The consumer node uses the content name and its selectors (the content attributes, resolutions, filters, etc.) in the interest message to request the content object. When any node receives the message, it first searches the desired content in the CS using longest prefix matching based on the content name. If there more than one content matches the content name, the selectors are used to precisely identify the content object.

3.2.1 Content

The terms "data" and "content" are used analogously in the context of this book. Any content that is desirable for the application, we call a "content object" or "data." Therefore, we will not go into more detail about the differences between data, content, and information. There are so many resources available online that differentiate between these terms [16] as follows:

- Data are the facts and figures known to be or assumed as facts that do no convey any meaning on their own.
- Content is a way to contextualize data. Date can become a content when they are presented in a usable form envisioned for one or more resolutions.
- Information is processed data that conveys the proper meaning or, simply put, information is data with meaning.

As mentioned previously, the terms "data" or "content," in the context of this writing, refers to the files or streams published by a producer or publisher. Therefore, the data or content entity can be a text document, video clip, audio file, or stream of text, audio, or video information. The size of the file can range from a few bytes to multiple gigabytes. These files are also known as "binary large objects" (BLOBs). In [17], it is stated that a large file or document is divided into multiple content objects and that the size of a content object can be up to 64 kB. Similarly, the maximum transmission unit (MTU) size of the Ethernet is 1500 bytes, which is the maximum allowable size of information that can be sent in a single transmission unit or frame. Therefore, it is evident that the data file or document can be larger than the content object as well as the MTU. As a result, the data or content or content object is fragmented in such a way to fit the permissible size of the MTU. After successful reception of these MTUs, the receiver node reassembles the content objects to form the whole. Content fragmentation and reassembly is discussed in detail later in this chapter.

Because fragmentation is unavoidable when we communicate large contents, a different approach, called "manifest," is used to efficiently manage the fragmentation and reassembly of contents. Simply stated, the manifest is a structure to index

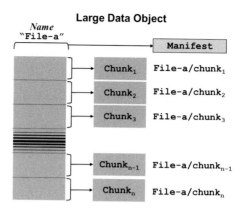

Fig. 3.7 Large data object with manifest

large contents and content objects. They represent the information as a metadata about the whole content object. The manifest by itself is considered as a content object and is sent through the data message with the payload of type "manifest," not as a "data." The following figure shows that a large content with the name "file-a" is divided into "n" chunks [18], and each chunk is named as "file-a/chunk$_i$." The manifest is maintained in this file, and it records the information regarding the type of content, access rights, publisher information, number of chunks, the size of the chunks, names, chunk-level hash-based names of the data objects, etc. [19]; refer to Fig. 3.7.

A manifest may also be used to represent the information about the stream data (either video, audio, text, etc.).

The manifests can also be considered the collection of linked content objects with corresponding hash digests. Large data objects are chunked, and each chunk shares the same data object name along with additional chunk information. Therefore, a manifest structure can be represented in the hierarchical structure, and the root of the manifest represents the content name. The root of the hierarchical manifest requires the cryptographic signature using the single public key for a whole data object and the subsequent chunks. In result, the large contents can be restricted to the size of the MTU, and thus fragmentation can be avoided. Due to this reason, a manifest can easily represent the represent the stream data (video, audio, text, etc.) as shown in Fig. 3.8.

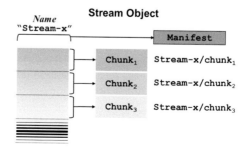

Fig. 3.8 Stream object with manifest

As discussed previously, the content object is divided into chunks, and CCN names every chunk. To secure a content object, either the whole content object is signed, or every chunk is signed individually. It is easy to sign a whole content object by signing the manifest of the content. Readers are suggested to read [19] for detailed information.

The content chunks are numbered and in a data packet, and each chunk is identified by its number. Along with that, the last chunk number field is also sent in the data packet to notify the maximum number of chunks into which the content is divided. These concepts are further explained in the following text.

3.2.2 Naming

The future communication architecture CCN uses content names instead of the host ID (IP address) to identify the content, and the name is also used to forward contents on the network. This means that every CCN-enabled node operates on the content names [20]. Here we focus on the content naming used in the CCN and its variant implementation in NDN. In CCN, the names are also termed "HSVLIs" (hierarchically structured variable length identifiers).

CCN and NDN use a human-readable naming scheme that uses the URL (uniform resource locator) like a hierarchical structure to name to the content. For example, the image content "safdar.jpg" of a Mobile and Network Laboratory (MoNeT) member in Kyungpook National University (KNU), Korea, can be represented as follows:

URL: **monet.knu.ac.kr/images/people/safdar.jpg**
Name: **/knu/monet/images/people/safdar/jpg/date/20151131/res/500/800**

Multiple segments in the URL provided above are used by the current communication protocols. The same content represented by the URL is also presented with the corresponding name used by the CCN. Each segment has been divided into multiple components, and the adjacent name segment components are separated by the delimiter "/". The content name is divided into different parts that are set and provided by the content provider. Unlike IP addresses, the length of the name components, as well as the number of components in the name, is arbitrary. The CCN protocol only uses the content name and its hierarchical structure to forward messages in the network.

The content name combines a routable prefix with an arbitrary suffix assigned by the publisher to a content or a piece of content. The routable prefixes and the suffixes do not have a fixed length. The routable prefix, e.g., "**/knu/Monet/**," is assigned the outgoing face and will be used to forward the interest message containing the content names with similar prefixes. Figure 3.9 shows the example content name divided into different segments that include the following:

Fig. 3.9 Content name used in CCN

(a) The globally routable segment, which is the network-wide coordinated name space, i.e., domain name. This name segment is used to route messages in the CCN.

(b) The content name assigned by the application (the applications can assign their contents a name in their own way). This application content-naming convention can vary and is application dependent so that each application can use any scheme.

(c) The last part of the content-name segment provides the components or attributes related to the contents used by the protocol. These name components include content-version information, chunk number, date and time stamp, and other content features.

The CCN protocol does not infer the meaning of the name but only it matches the name within the routing and related data structures. Therefore, the name components are set, assigned, and used by the application, and only the application can infer the purpose of each name component as in the previous example. The name components represent the application, organization, and global resolution, and they are reflected in forwarding rules. In CCN, the required content names can be provided by users (simply by copy-and-paste method or typing the name), a document or Web page in the form of links, generated by search engines, provided by the near field communication (NFC), or scanned from QR (quick response) code, or it can be provided in other formats.

Each content or every chunk of the content, collection of contents, or collection of chunks of contents is represented by the name. The name part that commonly represents the collection of the contents or chunks of contents is called the "name prefix." Therefore, the content names are also called "name prefixes" or simply "prefixes." The name prefixes are used to forward messages as well as find the contents by matching these name prefixes.

The advantages and disadvantages of hierarchical naming are briefly discussed in [21]. Hierarchical content naming has many advantages including, but not limited to, the following:

- The similarity of the content name with the current URL structure makes is easy to use it with most of the URL-based applications.
- The hierarchical structure of the content name prefix can easily aggregate different content names with similar prefixes under a single name prefix. For example, consider the following content names:
  /knu/monet/images/people/**safdar/jpg**/date/20151131/dim/500/700

/knu/monet/images/people/**hassan/jpg**/dim/430/620/color/RGB
/knu/monet/images/people/**azfar/jpg**/dim/300/403/res/180dpi
can all be aggregated under the name prefix:
/knu/monet/images/people/.
Instead of saving all the content names in the forwarding table, only the
aggregated name prefixes are saved, which minimizes the size of the
name-prefix table.
- Due to the small size of the forwarding table, the lookup operation can be
  accelerated.
- The lookup operation used for searching a name prefix from the name tables is
  similar to the one used in current routing devices. One of the most popular
  methods used for the lookup is the longest prefix matching (LPM) method.

Along with many advantages, the hierarchical naming scheme comes with a few
shortcomings [21]. The most prominent problem with hierarchical naming is that it
has variable size because the name is composed of a series of components with
variable number and length. This will result in large forwarding tables and slow
lookup process [22]. CCN-enabled applications may be interested in the content
without knowing the full name of the content. Searching a name prefix from the
table solely based on the content name is quite difficult and currently not supported
by the CCN. Along with that, the forwarding tables store only the prefixes and have
no information about the aggregated, which may cause the "suffix hole" problem in
CCN forwarding. For example, the prefix "/knu/monet/images/" has only a few
image contents that are available at the node or outgoing interface associated with
that prefix. However, based on the prefix information, all requests are forwarded to
that face even if the content is not available at that face. The CCN must use
additional information, i.e., content attributes, to search the specific prefix and the
appropriate content. For more information regarding security and other related
information about CCN-naming schemes, readers are advised to refer [20, 21].

3.2.3 Interest

It was clearly mentioned previously that CCN uses two types of messages: interest
messages and data messages (referred as a ContentObject message in CCNx documentation.). Here we will briefly discuss the *interest* message and its format as
well as how it is forwarded, i.e., what steps are taken by each node to forward the
message in CCN.

An interest message is generated by a node that requires any content, i.e., a
content-requesting node, called the "consumer node." The interest message uses the
content name to request and specifically identify the content or its chunks. Along
with the name prefixes, the interest message may also provide some specifiers or
selectors related to the content object that help to ascertain the explicit data in the
repository.

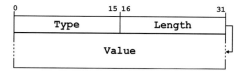

Fig. 3.10 Type–length–value format

CCN defines the detailed description of the interest message in the CCN message semantics [23] and CCN message type length value (TLV) [24] documents as shown in Fig. 3.10. Here we summarize the interest message in the TLV encoding and TLV fields in detail.

Basically, CCN uses **type** and **length** fields, both 16 bits in size, to encode the TLV packet format. The 16-bit size of a **type** field gives 2^{16} possible types, and the range of types from 0×1000 to $0 \times 1FFF$ is reserved for experimental use, but it provides ample space for the protocol types. Similarly, the **length** field defines the size of the **value** field (in octets) according to the **type** and **length** fields, and it can be up to 2^{16} bytes (64 KiB). The length value does not include the 4 bytes of the **type** and **length** fields. It is stated in [24] that a 0 value of the **length** field is permissible. The TLV is hierarchical in nature, and the **value** field may also contain other TLV structures (called "member" or "subordinate" TLV structures); that TLV is called the "container TLV."

The complete CCN packet format is shown in Fig. 3.11. Every CCN packet consists of a fixed 8-byte static header, which is common to all CCN packets and does not follow the TLV format. This common header is followed by the TLV(s) that represent the optional hop-by-hop header and packet payload (which is a CCN message) encoded in TLV format. There is an optional CCN message-validation TLV at the end of the packet.

The **packet payload TLV** is the CCN message that is an interest, data, or any other type of message used by CCN. The **version** field represents the packet version, and **version** = 1 at the time of writing this book. The **PacketType** field identifies that the packet is either 0 = interest, 1 = data, or other packet type (i.e., 2 = interest return, etc.). The **PayloadLength/PacketLength** is the complete packet length (in octets) starting from the **Version** field to the end of the packet. The **HeaderLength** represents the total length (in octets) of all the headers (including the fixed header) before the Packet Payload TLV or CCN message TLV. The **PacketType-dependent field** is divided into multiple other fields based on the packet type.

The fixed header is followed by the optional header or optional hop-by-hop header TLVs. The hop-by-hop headers are used in a CCN packet that requires per-hop (interest or data message) information propagation and/or operations.

Interest LifeTime TLV: As evident from the name, this header may be included in the interest message. The header represents the duration that an unsatisfied interest should be kept in the PIT. If this value is not used, i.e., not propagated with the interest message, then the default period that an interest must be kept in the PIT is

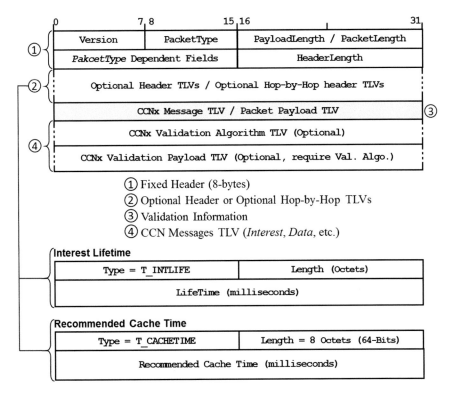

Fig. 3.11 CCN complete packet format

4 s. The **LifeTime** value is measured in milliseconds. It is also stated that the interest **LifeTime** header with **LifeTime** = 0 ms (**%×00**) should be forwarded; however, no data message reply is expected for that kind of interest message.

Recommended Cache Time TLV: This header is included in the data message to represent the suitable **LifeTime** of the content or data in the cache or CS. This value is recommended by the content producer or by any of the upstream data-forwarder nodes. The recommended cache time value is represented by the millisecond timestamp. Because it is an optional TLV, the nodes may ignore the suggested period.

(a) Interest Message Format

The complete CCNx interest message is shown in Fig. 3.12. The CCNx interest message TLV area, as well as the fixed header fields related to the interest message, are shown in the shaded area. **HopLimit** in the fixed header sets the maximum number of hops that the interest message can be forwarded in the CCN. The HopLimit value is set by the consumer node and is decremented at each hop. The interest message is forwarded until the **HopLimit** does not reach zero after the decrement. The maximum value of the HopLimit is $2^8 = 255$. Some of the

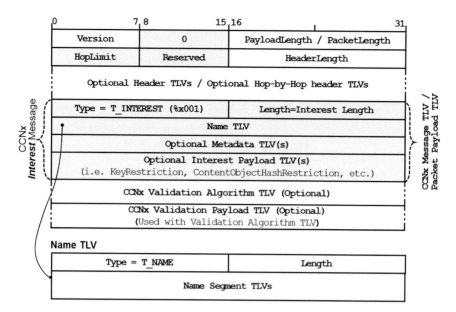

Fig. 3.12 Interest message TLV

Reserved field bits are used for the interest flags; however, there is no description of the flags used in the Internet draft [24].

The interest message container TLV starts with **Type = T_INTEREST** or is encoded as **%×001** (interest message type). The interest message TLV must contain the **Name TLV,** and a similar TLV is mandatory for the data message as well (the data message is discussed later in the chapter). There are other optional TLVs that can be included in the interest message.

(a) **Name TLV**: The name TLV container encodes the content name segment, and each segment of the naming convention is represented with a separate TLV and is assigned different **Type** values Refer [25] for more information on the Labeled Content Information document plus Fig. 3.13.

A generic name segment, which includes the arbitrary octets (hierarchical naming convention) is represented by the name segment TLV. The interest payload ID is used in the name segment to properly multiplex the interests based on the name with different payloads. The payload ID is created by the consumer node to identify the payload in the interest message. The interest payload ID is commonly represented by the hash of payload, which requires the hash algorithm information as well in the interest message. Another type of interest payload ID, called **NONCE**, is the name segment that uniquely identifies the interest message.

The application-component TLV shows the application-oriented payload in the name segment. It is suggested that the application name should be identified in the name segment and that the semantics of these components are application

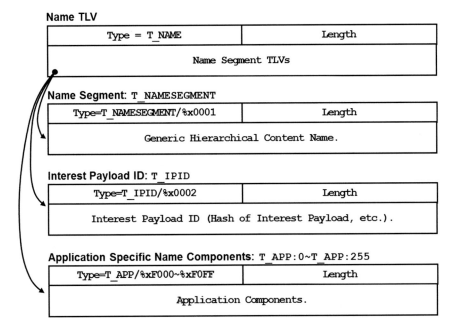

Fig. 3.13 Various formats of name TLVs

dependent. The name TLV is common in the interest and data messages; therefore, the same name TLV is used in the data message as well.

(b) **Interest-Specific TLVs**

These TLVs are optional and specific to the message type, i.e., interest message or data message. The metadata TLV(s) and interest payload TLVs are optional and are used with both interest messages and data messages. Here we will discuss the interest-specific optional TLV(s); the data message related TLV(s) are discussed later in the chapter.

At the time of writing this book, only two optional metadata TLVs related to the interest message are used to assist in identifying the closest matching content required by the consumer node. The interest-specific optional TLVs are the **KeyIdRestriction** and **ContentObjectHashRestriction** (refer to Fig. 3.14).

The key identifier is the unique string that offers methods to identify certificates that contains the particular key. The identifier can be a 160-bit SHA-256 algorithm hash of the key that is used to satisfy the interest or any unique number-generation mechanism. The **KeyIdRestriction** TLV sends a byte string that identifies the publisher signing key that will satisfy the interest carrying this TLV. The **ContentObjectHashRestriction** is the hash of the content object computed using an SHA-256 algorithm. This content object is the one that satisfies the interest message.

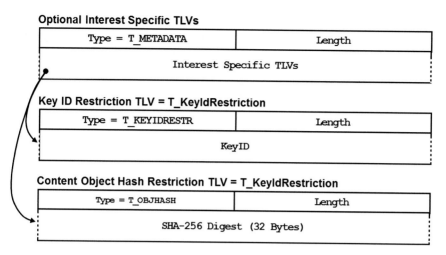

Fig. 3.14 Interest-specific TLVs

The validation algorithm TLV container comprises different TLVs that contain information about the specific algorithm used to validate the interest message and can be a cyclic redundancy check (CRC), message signature, SHA algorithm, or any cryptographic algorithm. The validation information, i.e., digest, CRC, signature, etc., of the interest is also sent in the nested TLVs of the interest message. It is important to note that the message validation is optional as depicted in Fig. 3.12. The next section briefly discusses the steps that are followed to forward an interest message.

3.2.4 Interest Forwarding (Interest Message Forwarding)

CCN uses pull-based or receiver-driven communication. The consumer node requires content to generate an interest message and transmits it in the network. The interest message is forwarded in the network until the required content is found or until the **LifeTime** of the interest message expires. The interest message is forwarded in the upstream direction, that is, from the consumer node toward the content-provider node.

Any node that has the requested or interested content that matches the specifications mentioned in the interest message replies with the content or ContentObject in the data message (it sends either whole content, if the content fits the permissible size of the data message, or a chunk of the content of the data message size in case of large content.). An interest message is satisfied with the content if the name in the interest message precisely matches the content name. Further, if the security is enabled, then the **KeyID** in the optional **Validation Algorithm** TLV of the data message should match with the **KeyID** in the **KeyIdRestriction** TLV of interest message. Moreover, the optional **ContentObjectHash** must be similar to the Content Object hash restriction within interest message.

Before discussing the interest message processing and forwarding steps, it is necessary to discuss the concept of local and remote next hops. In [17], the authors define that an application running on the node is the local next hop, and the remote next hop is the node that is not local to the current system. The concept can be easily understood by considering the **HopLimit** parameter in the CCN messages. If an interest with **HopLimit** = 0 is sent by the application running on the consumer node, then it can only be sent to the other applications running locally on the same node. However, if an interest with **HopLimit** = *m* is sent by the local application, then the message will be transferred and processed by the nodes that are *m*-hops away from the sender node. Similarly, when a node receives an interest with **HopLimit** = **1**, it decrements it first (**HopLimit** = **0**) and then forwards it to the local applications.

The stepwise processing and forwarding of the received interest message is discussed below:

1. Interest received from local applications (**HopLimit** = 0): Forwarded to the other applications running on the same node.
2. Interest received from the remote or downstream node with **HopLimit** ≥ 1:

   (a) Decrement the **HopLimit**.
   (b) If the resulting **HopLimit** = 0, then forward to local applications.
   (c) If the resulting **HopLimit** ≥ 1:

       (i)   Search the requested content name (and other optional search constrains if provided in the interest message) in the content store. If any match is found in the CS, then it generates a data message that carries the content or chunk of the content. The interest message is then discarded.
       (ii) If no matching content is available in the CS, then search the PIT. In the case that the node has already received the interest message requesting the same content, but with a different receiving face or with a different NONCE or interest payload ID, then the additional face or NONCE information is added in the previous PIT entry. However, discarding or re-forwarding the received interest message depends on the **LifeTime** value in the interest message and PIT record. The CCN suggests that the interest is forwarded only if the **LifeTime** of the previous entry in the PIT is less than the **LifeTime** in the received interest message. In this case, the interest message is forwarded when the current PIT entry **LifeTime** expires and the remaining **LifeTime** of the interest is set in the PIT entry. In contrast, the interest message is discarded and no further processing is required when the **LifeTime** of the received interest is less than or similar to the PIT record.
       (iii) In the case of no record found in the PIT for the received interest, a new PIT entry is created with a **LifeTime** similar to the one that is set in the interest message. Move to the next step (step iv).

(iv) Search name prefix in the FIB:

- If the FIB search returns more than one next hops, then the interest is forwarded based on the forwarding strategy [17], i.e., send the interest to all, best path, or alternatively to the paths based on their ranks, etc.
- If no FIB entry is found, then CCN holds the entry for a short time before discarding the interest message because the new FIB entry creation entry may provide a way to satisfy the interest. This is the crude functionality of the CCN, which requires more research.

(v) Send interest toward the upstream.

The reliability of the interest message is not presumed by the CCN, however, the reliability can be supported by retransmission of the interest message (either by the consumer node or the intermediate forwarder node) once the **LifeTime** of the message expires [26]. The total number of retransmissions and **LifeTimes** of the interest message should be recorded with each pending entry in PIT to support constrained retransmission of the interest message. In CCN, the default value of **LifeTime** is 4 s, and there is no specific value for the number of interest message retries. Detailed illustration of interest-message processing is depicted in Fig. 3.15.

3.2.5 Data Retrieval

The interest message pulls the desired content from the network using the content name (and the optional additional qualifiers to specifically identify the content or chunk of the content.) The content, either whole or chunk, is transferred by CCN protocol in data message(s). Here we will discuss the data-message format as well as the steps to successfully forward the data message toward the consumer node in the *downstream* direction.

The data message carries content as the payload, the name that identifies it, and additional information to verify and validate the content and the data message. This additional information includes the signature (cryptographic), the identification information of the publisher or signer, etc. Simply, the data message binds the *content* (may be chunk), its name, and the publisher, all together (refer to Fig. 3.16). It is required that all communicated data in CCN is verified through the signature and that every data message must contain the valid signature. An unverified data message should be discarded by the receiving node. Data verification requires signature verification from the signing authority using the public key; therefore, the public key should either be in the interest or included in the data message. However, in CCN there is no specific protocol for key distribution, and the keys are considered and distributed as the contents are.

The data message TLV consists of similar TLVs (e.g., **Version**, **PayloadLength**, **Name** TLV, etc.) similar to the interest message (refer to Sect. 3.2.3). The only difference is the **Packet Type** field, which identifies that the packet is *data = 1* packet, and there are separate data message specific TLVs. At the

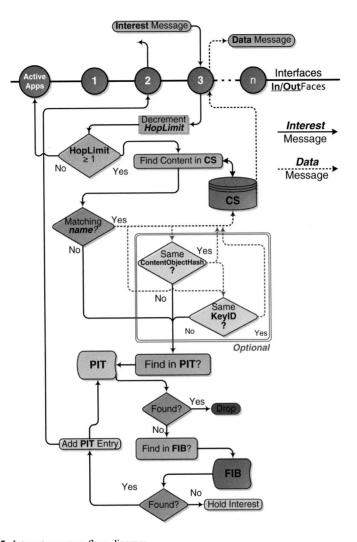

Fig. 3.15 Interest message flow diagram

time of writing this book, there are only two optional data message specific TLVs: called **PayloadType** (%×0005) and **ExpiryTime** (%×0006).

The **PayloadType** TLV defines the general type of the contents in optional **Data Payload** TLV, which follows this TLV. The broad types of the content types include the following:

- %×00: Data/**Content Object**: This is the default payload type when the **PayloadType** TLV is not included in the packet. Its contents represent the content object [complete or chunk of content] bytes used by the application.
- %×01: **Key** contents: Payload contains the encoded public key.

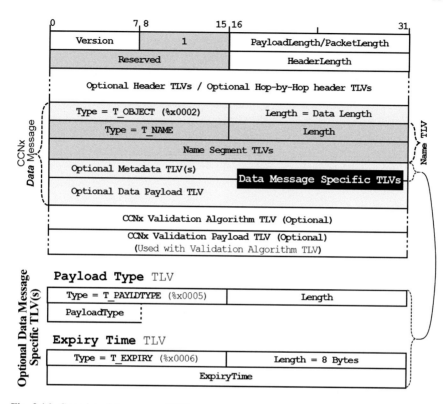

Fig. 3.16 Complete data message TLV

- %×02: **Link** information: This indicates that payload contains the link that is the tuple: {CCNx Name, KeyId, ContentObjectHash}.
- %×03: **Manifest** contents: The payload contains the *manifest* of the content object.

ExpiryTime TLV indicates an unsigned 64-bit integer that shows the time instance (encoded in the UNIX timestamp containing the number of milliseconds) when the payload will expire. The content provider should not satisfy the interest with a content object that has passed its expiry time instance. Content in the data message with no **ExpiryTime** TLV can be kept by the consumer or caching node as long as needed. The data message forwarder should not check the **ExpiryTime** field. The **Data payload** TLV contains the contents of the data packets.

Data Forwarding

When a data message is received, the following steps occur:

1. The node first searches all the related entries in the PIT by matching the information from the data message, i.e., content name, content hash, etc.

(a) If PIT search returns one or more entries and there is no **ContentObjectHash** restriction requirement, then the data message is sent to the recorded incoming faces. In the case of a **ContentObjectHash** restriction requirement, the **ContentObjectHash** in the data message is matched with the content hash in the PIT. *The* data message is transferred to the incoming faces recorded in the PIT with matching entries.

(b) When a **ContentObjectHash** restriction is requested in the interest message, the content hash in the data message should be checked in the PIT. In case of multiple PIT entries returned by the search with matching **ContentObjectHash**, the data message is forwarded to those faces that are associated with the PIT entries. Otherwise, the data message is discarded.

2. If PIT search does not return any entries, then the data message is dropped.
3. After transferring data message downstream toward the next hop nodes connected with the faces associated with PIT, the PIT entries are deleted. PIT entries with expired **LifeTime** are also purged.
4. Content received in the data message may be stored in the CS based on the caching policy [27], i.e., all caching, no caching, popularity-based caching, etc.

Detailed processing steps of a data message are depicted in Fig. 3.17.

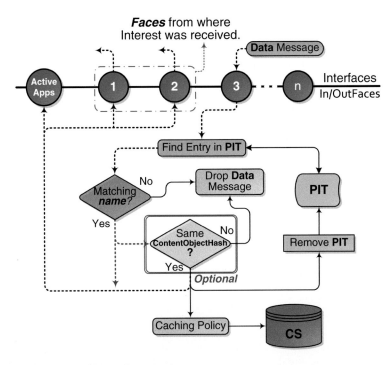

Fig. 3.17 Data-message flow diagram

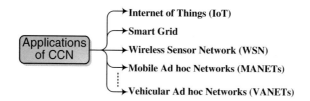

Fig. 3.18 A few examples of CCN-application areas

3.3 Applications for CCN

CCN is one of the promising future Internet architectures, and this is the reason that CCN has been actively investigated in many applications. Following is the list of applications where CCN has been investigated including, but not limited to, Internet of Things (IoT), Smart Grid, wireless-sensor networks (WSN), vehicular ad hoc networks (VANETs), etc. Figure 3.18 show a few of the CCN applications that are covered in this book.

3.3.1 CCN in IoT

IoT is the collection of small battery-operated devices having sensing, computation, and communication capabilities attached to the objects (anything) that connect them to the Internet. The attachment of the devices makes the objects "smart" to exchange and consume information to/from other devices autonomously or with little human intervention. Instead of communication in a point-to-point scenario, this huge network with a plethora of devices mainly focuses on data and information. Hence, the smart applications for IoT require contextual information that is generated either proactively or reactively by these devices [28]. There are several issues with these types of networks including, but not limited to, the following [29]:

- Dynamically addressing and naming each device
- Energy efficiency algorithms and protocols
- Network management and self-organization
- Network scalability and interoperability standards
- Network and service discovery
- Cloud connectivity and computation
- Real-time communication with small latencies
- Dynamic network partitioning and merging
- Mass data mining, filtering, and processing
- Scalable security solutions
- Mostly push-based communication
- Security, privacy, and trust technologies, etc.

IoT applications have been envisioned and realized in many applications as follows [30]:

- Smart healthcare system: anytime, anywhere person monitoring; remote healthcare; vital sign sensing, monitoring, and communication; emergency rescue and response; smart implants; telemedicine, etc.
- Intelligent transportation: connectivity and communication between vehicles, ships, trains, road signs, intelligent roads, tolls, planes, traffic lights, etc.
- Building and infrastructure interconnectivity and monitoring.
- Intelligent home.
- Connected smart city.
- Smart grid and energy sources.
- Connected manufacturing and intelligent products.
- Cooperative sensing.
- Smart logistics and retail of goods, etc.

IoT has been actively researched due to its huge range of applicability. Therefore, the future communication architecture, CCN, is also investigated in the context of IoT, and the solutions are discussed here. In [31], the author proposed and implemented ICN architecture for IoT. IoT contents are addressed using the names and its benefits of implementing the concept in home-automation systems are highlighted. Its implementation uses push-based communication with the home-automation system.

In [6], the authors used a push-based CCN-communication mechanism for IoT traffic. Initially they described a subscription scenario where the subscription is performed by sending an interest message that uses the hierarchical name for the desired content without creating a PIT entry for each interest forwarded. The reason for not creating the PIT entry is that no data are immediately expected for this interest. However, the authors mention that PIT-entry creation will avoid the interest looping. The authors use a timestamp in each content name (the time stamp is stored in the FIB as a separate field, called "last_seen") to avoid the routing loops. Each time a content is forwarded, its timestamp is matched with this field in FIB, and only the most recent data are forward to the subscribed consumers. To avoid redundant forwarding of the regularly sampled data, an optimal forwarding strategy is also proposed. In the optimal forwarding strategy, the provider pushes data only for the requested time period that is demanded by the consumer (the consumer subscribes for the data sampled in the specific period.). The intermediate nodes keep track of this dedicated period using the counter. Data messages with similar samples and their overlapping counters or sampling periods are combined and forwarded as a single data message instead of separate messages. In the end, a smart-pushing strategy is proposed to limit the number of forwarded messages. Identical-interval subscriptions uses one counter, and the remaining subscription intervals consider separate interval counters.

Another proposal that analyzes the application of CCN in the context of IoT is presented in [32]. It suggests that the complete ICN mechanism cannot be implemented on IoT nodes because of their power, sensing, processing, and memory constraints. Therefore, some functionalities, i.e., security and caching options, are delegated to the third-party trusted nodes. Optimal caching can be achieved by only storing the latest sensed information obtained using the counter or sequence numbers. Because data from different network segments can be combined, the use of a sequence number in data naming is not a valid solution. The authors tackle this issue by using the lifetime parameter, called "FreshnessSeconds," in data packets, which indicates the period until which data can be held in the CS. The authors simulated the IoT environment to check energy and consumption, as well as bandwidth use for varying CS sizes, and compared it with IP-based communication.

The authors in [33] performed experiments with an NDN implementation on IoT deployment in multiple office buildings. The bare-bones open-source Linux version of the CCN, called CCN-Lite [34], is ported over the RIOT [35], which is an operating system for resource-constrained IoT devices. The implementation uses the hierarchical naming scheme, which is better suited for routing aggregation. In the experiment, a simplest interest-flooding mechanism, named "vanilla interest flooding" (VIF), is used where each node forwards the interest in the network when it receives the interest for the first time. There is only one producer and many consumers in the network. The number of radio transmissions of interest messages are reduced, and the dynamic FIB is managed by the reactive optimistic name-based routing (RONR). RONR assumes that the FIB is empty at the network initialization; then first interest is flooded in the network, and the name prefix of that interest is stored in the FIB. The subsequent interests for the chunks of the same content are auto-configured in the FIB and forwarded in a unicast manner based on the name prefix in the FIB. It is assumed that the whole content is available at a single provider, which may not be the general case. The FIB entries are timed out to minimize its size. The authors also analyzed the impact of caching versus no caching with single or multiple consumers.

Another CCN proposal for IoT was proposed in [36] that uses NDN for an IoT scenario with multiple data sources located at single hop, single interest, multiple data (SIMD). They proposed a collision-avoidance mechanism by introducing the old concept of the contention window. Multiple data sources simply select random number CW_r between [0, $CW_{max} - 1$]. The data source defers its transmission until the collision-avoidance random period $\tau_{ca} = V_{slot} * CW_r$. A new interest packet is introduced that requests data from multiple sources, multisource interest (msINT). The consumer advertises V_{slot} and CW_{max} in msINT. The CW_{max} is dynamically adjusted based on the number of the data sources. The simulations are performed to measure the number of interest, interest overhead, and collection time, and compares it with the single interest, single data (SISD) scenario.

Most of the IoT proposals just focused on the simple scenario where an interest is used to subscribe for data, and the data are sent for that subscription period. Generally, there is periodic sensing in the IoT application, but the data may also be generated in response to the event detected by the sensors: this is called an "unsolicited emergency notification." The application requires this notification to be pushed into the network in either a unicast or multicast manner. However, there should be CCN proposals that implement the unsolicited emergency-message communication support along with the solicited data communication in IoT. There are several other issues that must be addressed when adopting CCN in IoT, and the readers are suggested to explicitly refer the related work.

3.3.2 CCN in Smart Grid

It is quite hard to define the term "smart grid"; however, we define it as follows: *Smart grid integrates advanced communication, control and automation, computer-based technology and systems that manages, regulates, and brings the responsive and resilient utility electricity network.* The grid connects the power-generation sources and manages the electricity demand in a reliable, sustainable, and economic manner. Balanced demand-based supply of electricity is one of the main objectives of the smart grid because most of the electricity generation relies on fossil fuels, which increases the amount of harmful gases in the environment.

The smart grid is a system of systems consisting many components including, but not limited to, the following:

- **Smart meters**: Utility meters are enabled with communication technology to connect energy consumers and providers to automate billing, regulate demand, and detect faults for speedy recovery.
- **Smart electricity generation**: Several power-generation sources ranging from renewable to ones that consume fossil fuels. A smart electricity-generation system optimally generates electricity to meet users' demand with minimum cost and carbon emission.
- **Smart power distribution**: A power-distribution system connects the power-generation sources with the consumers through distribution lines and smart substations. It has self-optimizing, self-healing, and self-balancing capabilities to automatically predict and detect power failures in real time.
- **Smart substations**: These control and monitor critical and noncritical operational data, i.e., battery status, transformer status, breaker information, power-factor performance, security, etc.
- **Information and communication technologies**: These provide means for all the components to interact with each other to conserve energy by efficiently using, distributing, and generating electricity.

The rest of the components are shown in Fig. 3.19, but further details on the topic are out of the scope of this book.

Recently ICN architectures and their effectiveness has been investigated in smart-grid systems to investigate its feasibility and effectiveness. These proposals are summarized below.

The authors in [37] promote the use of ICN in smart-grid applications and suggest the use of publish/subscribe, such as communication, to ease the smart-grid control with simple and secure data sharing. The smart grid also uses a

Fig. 3.19 Components of a smart grid

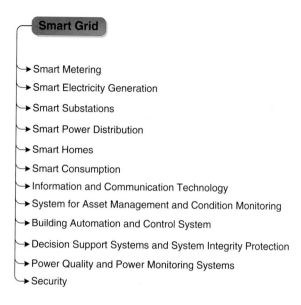

many-to-many data-communication approach between devices and applications; therefore, ICN is envisioned to be a proper communication architecture for smart grids in the future. Currently, the ICN framework is used as an overlay on smart-grid communication to enable seamless and robust communication.

CCN-based advanced metering infrastructure (CCN-AMI) for the smart grid was proposed in [38]. The main objective of the proposal was aimed to efficiently control communication congestion, support mobility, and ensure security. The authors explicitly discussed the naming scheme for the home network, which is an integral part of AMI. The naming scheme consists of several components—including access scope (service, device, content, policy, etc.), transmission mode, and message type—along with version and segmentation information. In addition to the naming scheme, the authors briefly discussed the components encompassed by the CCN-AMI architecture: smart meter, customer energy-management system, smart-home appliances, data connector, meter data management system, demand response management system, load-management system, etc. The authors evaluated the performance of CCN-AMI with the current IP-based smart-metering infrastructure. The results show that CCN-AMI can significantly reduce network congestion traffic. The same authors proposed a key-management scheme for the CCN-AMI in [39].

In [40], the authors claim they implemented the ICN-based communication infrastructure, called "C-DAX," to support data communication in a smart grid. They proposed the C-DAX architecture and its components and planned to have a fully functional laboratory demonstration as well as port the implementation of the actual smart-grid system in the Netherlands.

3.3.3 CCN in WSN

Wireless-sensor networks (WSN), an integral part of IoT, are the collection of a large number of small battery-operated devices that have sensing and communication capabilities. The WSN consists of inexpensive and large number of devices spread or installed in the sensing area that monitor environmental or physical parameters, i.e., humidity, temperature, pressure, vital signs, water salinity, soil moisture, etc. A typical tiny sensor node consists of the following components connected as a single component:

- Communication
- Computing
- Sensing
- Power source
- Actuation

The sensed parameters are communicated wirelessly in an ad hoc mode toward the collection point(s), called "sink node," either periodically or based on an event for further decision making. The sink node(s) can be static or mobile to collect data from the WSN. The main objective of WSN is to monitor the field for a very long duration, and that is only possible by conserving the battery power [41]. The operations that consume battery power are wireless communication, sensing, processing and listening. Therefore, a node must efficiently schedule its operations. Along with the longer lifetime, the WSN must self-configure and self-organize due to dynamic network architecture caused by node failures [42]. Several routing solutions and proposals have been presented in the past to achieve energy efficiency, self-configuration, and self-organization. Researchers in the area of WSN may further refer [43–45].

Recently CCN communication architecture was investigated in WSN [46–51]. In [46], Bilel Saadallah et al. implemented the CCN named-data communication stack in the Contiki [contiki] operating system. *Contiki* is one of the operating systems for the wireless-sensor networks and embedded systems that contain resource-constrained devices. The implementation considers hierarchical naming (similar to CCN) consisting of the name prefix followed by the content attributes as follows:

Prefix	**/Temperature/halinria**
Content Attributes	/Temperature/halinria/**Bld1/Floor2/Office1**
	/Temperature/halinria/**Bld1/Floor1/Office12**

The implementation uses interest and data messages of 102 bytes to match the IEEE 802.15.4 frame, which is 127-bytes long (127 to 25 (-bytes MAC header) = 102 bytes). The processing steps of these messages are modified to suit the processing capabilities of the WNS nodes. CCN PIT, FIB, and CS are also implemented accordingly. CCN implementation in Contiki is evaluated through simulations and real deployment using a synthetic monitoring application for varying network sizes.

The authors in [47] implemented the CCN-communication architecture for WSN in Wieslib [52], which is a library of algorithms that form heterogeneous sensor networks. CCN implementation for WSN is termed "CCN-WSN" and implements hierarchical naming, interest message, data message, PIT, CS, and FIB. Evaluation of this flexible implementation of CCN-WSN demonstrates the suitability of CCN in WSN. The CCN architecture for WSN was proposed in [48]. The architecture is divided into two tiers: The first tier manages the heterogeneity devices that comprise the WSN (sensor node, sink, remote server). CCN is enhanced with some changes to the forwarding strategies to improve data collection. The second tier is the modified, lightweight, and shortened CCN forwarding strategy encompassing forwarding data structures (FIB, CS, and PIT), message (interest and data) transmission, and techniques for message retransmission.

Amadeo et al. [49] explored the application of NDN ICN architecture in WSN. To increase the reliability of interest and data message communication, the authors proposed the defer-window concept for both types of messages where the defer window of the interest message is larger than the data message. Rebroadcast of data and interest is deferred for the duration of D_{DW} and I_{CW} duration, respectively, as follows:

$$D_{DW} = \text{rand}[0, DW] * \text{DeferSlotTime}$$

$$I_{DW} = (DW + \text{rand}[0, DW]) * \text{DeferSlotTime}$$

The DeferSlotTime is a short time and fix interval. The reason for smaller data DW than the interest message is to provide it the higher priority. The results of the proposed defer window based NDN implementation are compared with the basic NDN implementation in WSN. The results show that the proposed defer window based NDN architecture is more energy efficient than the basic NDN implementation in WSN. The CCN implementation of Contiki [53] has been simulated and implemented on TelosB nodes, and the results are discussed in [50]. Both the simulations and the implementation show that the proposed solution has less message exchange overhead with acceptable data-retrieval delays. The authors in [51] proposed the fine-grained NDN-based trust model, which allows both consumers and providers to authenticate sensed data and access control mechanisms based on data encryption and key attributes.

The previously discussed solutions comprise the preliminary CCN implementation and their validations in WSN. However, more serious attention from researchers is required to propose more energy efficient CCN-based solutions for WSNs.

3.3.4 CCN in MANETs

Mobile ad hoc network (MANET) is an autonomous, infrastructureless, self-healing, self-organizing, dynamic topology, and multi-hop network of mostly battery-operated nodes. These characteristics of the network make data delivery between source and destination node a more challenging. Due to the multi-hop nature and dynamic network topology, data dissemination with less control overhead conserves energy and is more reliable. There has been much research to resolve these issues; however, the issues still remain, and research is still ongoing. Much literature and many books are available on the topic, and readers can consult the following very recent articles to quickly obtain information about the topic [54–58].

After the invention of a new future emerging network paradigm, CCN, which substitutes the traditional IP-based, location-dependent, and host-centric networks with the name-based, location-independent, and data-centric network, researchers investigated CCN in MANETs. Following is the brief summary of CCN-based schemes for MANETs.

In [59], the authors proposed the NDN-based new data-forwarding scheme for MANETs termed "listen first, broadcast later" (LFBL). LFBL uses three-way message exchange (announce the name prefixes, forward the interest, and return the data), which is supported by NDN. Initially a node disseminates the network-wise request carrying the names of the data requested by the application. Any node with the requested named data sends a response packet backward to the requesting node. Next, the data-receiving node sends the acknowledgement (positive or negative) packet. LFBL saves the prefixes that are flooded with the requests. When an intermediate node receives the packet, it defers its transmission for a short time and listens to that channel for any potential forwarder or responder. The results show that LFBL has better data-delivery efficiency than the famous reactive routing protocol called "ad hoc on-demand distance vector" (AODV) running over 802.11MAC and CSMA MAC.

The authors in [60] implemented the CCN architecture on laptops running Linux OS to create on-demand MANET. The CCN-based protocol suite was implemented by the authors. The CCN data structures implemented includes CS, metadata registry, and an interest table. The large MANET for emergency and tactical scenarios with hierarchical architecture, group mobility, and operation are considered in the paper. The hierarchical storage and search mechanisms for the topic-based and spatiotemporal contents are considered in the implementation. Initially the publisher disseminated the same meta-content information that is disseminated through gateway nodes at the upper layer of the hierarchy in each group and recorded at each node's registry in each group. The node sends the interest to the gateway node, and the gateway replies with the matching content if it has the copy. Otherwise, the interest is sent to other gateways to find the content, which makes these gateways responsible to publish and deliver data.

In [61], the authors pointed out basic approaches to identify fundamental points in the design of content-centric MANET (CCM). The performance and efficiency of the identified design are evaluated through modeling. Announce content availability, send query to find the content, and fetch the content are three major operations considered in the modeling space. Three schemes are identified in the paper to retrieve content in the MANET. Reactive flooding describes the requesting node flooding the requested message to find the content, service, or location of the content or service-provider node. In this case, there is no announcement; a node just floods the query message, and the content provider replies with data or service in a unicast manner to the requesting node. The second proposed design is proactive flooding where each node periodically floods the resource or content available in MANET. The requesting node just listens to the periodic announcements, and then the request and content delivery are performed in a unicast manner. The third and the last CCN design for MANET uses geographic hash tables (GHT). This design assigns a key to each resource in the network and, when operated through this key, it may provide a pair of two-dimensional (x, y) coordinates. The node first announces the pairs (resource, host) to the nodes closest to the resource key. The requester first computes the hash of the resource in which it is interested, which provides the location of node(s) holding the pair of (resource, host). The query is forwarded in a unicast manner to these nodes through a GPSR protocol. The content fetches operation is performed once the resource host information is retrieved. Fetch and content forwarding are unicast operations. Content availability, latency, and overhead cost are modeled by the authors.

In [62], content centric fashion MANET (CHANET) has been proposed. The CHANET architecture provides detailed processing of the broadcast-nature interest and data messages, and the many features for 802.11-based MANET include the following:

- Hierarchical naming
- Content segmentation and reassembly (content divided into chunks)
- Content advertisement (periodically by fixed providers, e.g., access point)
- Content discovery (interest message) and delivery (data message)
- Retransmission request (int-ack message)

CHANET brings simplicity and robustness by using the broadcast nature of CCN messages. Nodes overhear the broadcasted messages and defer the time to reduce the number of collisions. The interest and retransmission request message forwarding decision is made by the node itself that receives the messages. It indirectly provides the retransmission request (embedding acknowledgments in the interest packets) and sequence control mechanism. The consumer provider mobility are inherently implemented and supported by the scheme.

The open-source CCN emulation for MANETs in [63] is called CCN-Joker "CCN-Java Open Source Kit EmulatoR." The emulation considers basic aspects of CCN including interest, data, cache-replacement policy, etc. The authors did not consider the CCNx (the implementation of CCN) because they wanted to have full control over the replacement policy and a lightweight application program that can easily be executed by resource-constrained devices. The main objective of selecting Java was due to its networking library (e.g., java.net) and multithreading capability. CCN-Joker implements FIB, PIT, and CS along with two additional structures called "repository" (permanent storage for content seed copies) and "my requests" (pending requests that are generated by the node itself). The implementation distinguishes between local and external interests. The interest and data messages are transferred to their respective handling modules called "request manager" and "content manager." The assignment of interest and data messages to their respective handling modules after they are received is performed by the Joker Server module. The operations of these modules are similar to CCN core implementation.

Adem et al. [64] proposed a CCN scheme for MANETs that avoids the message loss in MANETs. The scheme uses the neighborhood information and defer time to achieve this goal. Detection of neighbors is performed by periodic overhearing of the communication activities of the neighbors. If a node does not hear any periodic advertisement or any communication activity, the link to that neighbor is considered broken. The interest and data messages may be broadcasted or unicasted. In case of broadcast messages, the time to live (TTL) value is used; however, TTL is not considered for the unicast messages. If a node has any content, it is advertised by that node. For intermediate nodes that receive this content, advertisements may record all the paths toward the content source in the FIB. Each path entry has hop-count and may record any other path metrics. The FIB entries are sorted based on the hop-count. When an interest packet arrives, most of the operations are similar to CCN other than the FIB search and next-hop forwarding. The FIB entry with minimum hop-count is used to forward the interest message. The interest message is broadcasted when no alternative path is available. In the case of node mobility, the path loss is frequent. When a node has no entry in the FIB, then it calculates the defer time and starts overhearing the neighborhood communication. At the expiration of the deferred window period, the node checks the FIB for the possible next hop entry; otherwise, it broadcasts the interest message. It is the same case with the data message. To summarize the discussion, the proposed scheme ensures the next hop entry before using it and discovers the alternative path or means to forward the messages.

An interest flooding control scheme, called "neighborhood aware interest forwarding" (NAIF), for NDN enabled MANETs has been proposed in [65]. NAIF initially prunes the ineligible forwarders because it selects the potential forwarders from the forwarder set based on the data-retrieval rate and its distance to the content provider. Afterward, the eligible potential forwarder may probably drop the received interest based on the updated smoothed forwarding rate of the related name prefix. After every update interval, t, the forwarding rate (F_t) is smoothed as $F_t = s_{int}/r_{int}$ where s_{int} is the number of interests sent, and r_{int} is number of non-cached interests received. This information is used to adjust the interest-forwarding rate.

The authors in [66] proposed an enhanced version of CHANET [62] called "E-CHANET" (enhanced content-centric multi-hop wireless network). It is claimed that E-CHANET solves issues such as channel unreliability, dynamic network topology, broadcast storm, data message reliability, and interest rate control as well as achieves energy efficiency in wireless multi-hop ad hoc networks. The interest and data messages are extended with additional information, and a new data structure—along with FIB, PIT, and CS—called "CPT" (content provider table) is used to achieve the above-mentioned goals. The interest message piggybacks hop-count (*iHC*), provider node ID (*sProvID*), and expected hop distance to the selected provider (D_p). Similarly, the data message carries additional information, for instance, sustainable interest rate information, hop-count same as the interest message, provider node ID, and distance to the consumer. If a consumer has no previous information of the potential provider, then *sProvID* and D_p are set to null. When a node receives an interest with the matching content provider, it initiates random defer time before data transmission. If the requesting content is not found, then the node checks the PIT and adds an entry. Before forwarding the interest message, the node must check the CPT. The interest message may include the D_p and provider ID if the potential provider is found in the CPT. This node uses a counter-based suppression technique to schedule the interest forwarding. The provider node only replies to the first interest received by it and discards copies of the same interest received later. Each intermediate data message receiving node uses hop-count information to forward the data packet toward the consumer node. Each interest message is assigned a timer (retransmission time-out) to schedule the next retransmission of the interest message. If a consumer node fails to receive data after the specified number of tries, then the consumer gives up trying to obtain the data and may try later.

The authors in [67] implemented the content-centric design for MANETs one handheld device called "SCALE." The authors use nodeID to reflect its geographic coordinates, content name, and several other parameters to achieve content communication. The demonstration supports three functions: publish, subscribe, and query the content. The data in the cache is indexed based on the actual content instead of the content name.

A topology-aware CCN protocol (TOP-CCN) was proposed in [68] employing multi-point relays (MPRs), publisher MPR (PMPR), and a hop count based flooding-control mechanism. The PMPR announces multiple contents in a single announcement to reduce announcement overhead and reduce flooding range using the hop-count limit. The MPRs are used to discover and announce the contents, and every node periodically elects MPRs. These MPRs reduce the number of announcement and content-discovery broadcast messages. The topology information is stored in tables that include one-hop and two-hop neighbor node IDs and lifetimes. The content announcement message includes the content prefix, MAC address of the node, content prefix list of neighbors, neighbor IDs, type of neighbors, the TTL of announcement message, cache hit ratio, and flooding type (one-hop/multi-hop). This information is used to maintain a neighbor list as well as make MPR selection.

A topic-based publish/subscribe content-centric network (TPS-CCN) system for MANET was proposed in [69]. The proposed TPS-CCN system uses a hierarchical naming scheme, CCN functionality, in-network caching, and multicast functionality to minimize delay and maximize delivery efficiency. Through dynamic configuration of CCN-forwarding tables, the mobile device caches data and can function as a data "mule" to transport contents toward the disconnected parts of the network. This realizes the delay tolerance in CCN networks.

TPS-CCN generally performs four functionalities: publish contents, publisher discovery and sync, pull-based data communication, and delay-tolerant data communication support. A data publish procedure executing on a node makes information items available to the related TPS-CCN subscribers. Published and unsatisfied information is stored in the information item repository (IIR) and a pending interest repository (PIR), respectively. The interest messages are processed and forwarded based on these tables. The data-pull functionality is executed by the TPS-CCN subscriber to pull content from the publishers. In case of multiple publishers of the same topic, a separate window is maintained for each provider to speed up the transfer procedure in the presence of long round-trip times. To pull information content from the publisher, the subscriber requires the topic name, publisher ID, and the sequence number of the last produced content. The subscriber periodically executes the publisher discovery and the sync procedure to retrieve updated contents. When a subscriber becomes aware that it is disconnected from the publisher, the delay-tolerant mode is activated. Publisher disconnectivity can be detected by the publisher-discovery procedure. The DTN mode is well explained in [69]. The performance evaluation of the CCN-MANET routing engine was developed as a plug-in of the OLSR Linux daemon and executed under the emulated Linux virtual-machine environment.

The authors in [70] analyzed two NDN-forwarding strategies: the blind-forwarding or provider-blind forwarding strategy and the provider-aware forwarding strategy. The blind-forwarding strategy is a counter-based broadcasting scheme, which defers data and interest message transmissions to limit collision probability and controls interest redundancy by overhearing the messages. This scheme achieves the above-mentioned performance in a scenario where NDN messages are broadcasted without any neighborhood knowledge and identity of the content sources. In contrast, the provider-aware forwarding strategy uses the soft-state information of the content sources, i.e., the distance to content source, content source ID, etc., to facilitate content retrieval. This soft-state information about the content sources is maintained in table(s) at the nodes, such as a distance table, and this information is piggybacked in interest and data messages. The above discussed schemes are summarized in Table 3.1.

Because most of the MANET nodes are battery operated, mobile, and communicate without infrastructure support, the CCN solution must efficient in content discovery, message-forwarding reliability, and support security. The proposed solutions focus solely on the application and testing of CCN architecture, and more solutions are required to support the reliability and efficiency of CCN messages as well as alleviate bandwidth congestion resulted from the broadcast nature of the network.

Table 3.1 CCN-/NDN-based schemes for MAMETs

Name	Scheme	Architecture	Simulator	Comparison	MANET
LFBL [59]	Reliable forwarding	NDN	Qualnet	AODV over 802.11MAC and CSMA-MAC	General with RWP mobility
MANET CCN [60]	Metadata and content forwarding in hierarchical MANETs through gateways that manage the group	CCN	Implementation	OLSRD	Hierarchical architecture, group mobility
CCM [61]	Announce, query, and fetch operations; reactive and proactive flooding; hash table based design	CCN	Modeling latency, delay and availability of content.	–	General
CHANET [62]	Content-centric fashion MANET with retransmission support	CCN	NS-2	FTP over TCP/IP	802.11 with AP*
CCN-Joker [63]	Java-based fully customizable and open-source emulation of CCN-like architecture	CCN	Java based implementation	–	General
Adem et al. [64]	CCN message loss avoidance in dynamic MANETs using the defer-time scheme	CCN	OPNET	Typical CCN	General
NAIF [65]	Neighborhood-aware interest forwarding based on interest-forwarding rate	NDN	Qualnet	LFBL and NDN forwarding daemon	General
E-CHANET [66]	Enhanced CHANET with retransmission and congestion-control support	CCN	NS-2	CCN, CHANET, and TCP/IP	802.11 with AP*
SCALE [67]	Geographic location of device, content, name, and other parameters are used to implement content communication	ICN	Implementation on Android devices	–	Generic
TOP-CCN [68]	MPR-based content announcement-and-discovery protocol	CCN	NS-3	CCN and E-CHANET	Grid wireless network
TPS-CCN [69]	TPS-CCN publishes content, publisher discovery, and sync pull-based data communication and delay-tolerant data communication support	CCN	CCN MANET routing engine implemented as a plug-in of OLSR Linux daemon	UDP-Ack, no DTN, no cache, and reliable IP multicast	Sparse MANET
Amadeo et al. [70]	Blind forwarding and provider-aware forwarding	NDN	NS-3	–	Wireless network with AP* and MANET

*Access point

3.3.5 CCN in VANETs

Vehicular ad hoc networks (VANETs) are a reality of the near future and play vital role in everyday life by providing services, e.g., safety and comfort of passengers while driving, infotainment, etc. The network comprises vehicles with communication capabilities. One of the most prominent characteristic of the VANET is its highly dynamic topology, which makes it challenging to propose any efficient and promising communication solution. Along with that, it has also been proven by many researchers that the TCP/IP communication protocol stack is inefficient for mobile networks. This is the reason that a separate protocol stack for VANETs, called "wireless access in vehicular environments" (WAVE) [71], has been proposed. WAVE supports data exchange without the TCP/IP overhead through the WAVE short message protocol (WSMP), which was designed for safety critical and control messages.

Recently, there has been much research with the intention to reliably communicate emergency information, traffic status, vehicle sensory data, and infotainment information in a non host centric manner by using an information-centric communication mechanism. Content-centric approaches are briefly discussed and summarized in [72]. Here, we will discuss and summarize the very recent CCN-based schemes for VANETs.

The very first proposal in [73] employs a named data communication mechanism to collect vehicles' sensory data from the mobile vehicular network. This information may be used by the manufacturing or related companies to provide vehicle management, safety, and alert information to the drivers or owners of the vehicles or companies. A hierarchical naming scheme, with company, type of information, country, state, etc., is used to request this sensory information.

A CCN-based multi-interface enabled vehicular network has been proposed in [74] where each vehicle may be equipped with more than one network interface to communicate with other vehicles or infrastructure (e.g., 802.16 (WiMAX), IEEE 802.11p, WiMAX, and UMTS). It supports a general pull-based vehicular network data communication and push-based (unsolicited messages) safety-messages dissemination as well. The push-based packet is called the "event packet." It is possible that any vehicle can generate a denial of service (DoS) attach by consuming the network resource by sending a large number of event packets. This is controlled by limiting the number of such packets from the neighboring nodes. The format or structure of the event packet is similar to the CCN's ContentObject or data message.

CCN and NDN use a hierarchical naming scheme with general components to identify the contents. The naming scheme for vehicular networks to represent spatial and temporal vehicular information has been proposed in [vccn03]. This naming scheme is used by vehicular applications to communicate contents between vehicles and the infrastructure. The vehicular network information is identified by the following naming scheme: **/traffic/geographic_scope/temporal_scope/ data_type/NONCE.**

The CCN architecture was implemented in 802.11p-enabled vehicles and RSUs in [75]. It uses three types of messages: **B-Int**, **A-Int**, and **C-Obj**. The content name is used in all messages with the potentiality to discover, identify, and retrieve content or a chunk of the content. **The B-Int message** is broadcasted to discover the content provider including the content name, sequence number (to avoid duplicate), hop count, and additional options. On receiving the **B-Int**, a node searches its content store, and if it finds the matching content it sends the C-Obj (s) after time-out of the random defer time. In case of multiple copies of the **A-Int** message, only the first one is answered. Intermediate nodes that maintain **C-Obj** maintain entries about the content provider in CPT: provider ID (MAC Address), hop-count distance, and the provided content ID. The **C-Obj** message consists of the content name, provided content chunk information, segment information, hop-count, provider ID, and the content itself. After receiving the first content chunk, successive chunks are requested by sending the **A-Int**, which carries additional information compared with the **B-Int**: preferred provider ID, expected hop distance to provider, and acknowledgement of previously received chunks in a bitmap format. The simulations show that CRoWn has a better content-retrieval rate compared with the legacy TCP/IP-based communication protocol.

Authors in [76] proposed a content centric communication scheme for autonomous vehicles, named "CarSpeak." CarSpeak enabled autonomous vehicles communicate sensory information from neighboring cars as well as the sensor installed on the static infrastructures, e.g., RSU, road side buildings, etc. using the content-centric interest-data mechanism.

A content-centric vehicular network (CCVN) transport strategy has been proposed in [78] that is almost similar to the one proposed in [77]. A preliminary analysis of CCVN and its effectiveness was performed in [79]. The main difference compared with [77] is the consumer- and provider-driven handovers proposed in [78]. In the consumer-driven handover, the consumer selects the nearest provider based on the providers' information from the CPT. In the case of no provider information, the consumer sends a new **B-Int** message. If a node that is not enlisted in A-Int as a potential provider receives the **A-Int** message and also has the matching content in CS, it can reply with the **C-Obj**. This is called the provider-driven handover. A node only selects itself as a new potential provider if it has less hop distance than the provider in **A-Int** message.

Due to the broadcast nature of wireless faces, collision of interest and data messages is inevitable. The authors in [80] introduced different timers to avoid the NDN-message collisions in an NDN-enabled vehicle-to-vehicle (V2V) multi-hop highway communication network. The timers include a collision-avoidance timer, pushing timer, NDN-layer retransmission, and application retransmission timer. With the collision-avoidance timer, a node randomly delays the interest in data transmission between 0 and 2 ms. Push type messages are scheduled based on the pushing timer, which is computed based on the transmitter-receiver distance, maximum transmission range, and minimum next-hop delay. The NDN-layer retransmission (50 ms) and application retransmission timers are used to schedule the retransmission of NDN messages over the lossy wireless network.

Hierarchical Bloom-filter routing (HBFR) has been proposed for CCN-enabled VANET in [81]. HBFR identifies each chunk of the content object with a hierarchal name e.g., /category/service-name/additional-info/. The category shows the type of content based on its size, type, popularity, etc. The service-name identifies the data services provided by the node(s) based on time-sensitivity of the content. The additional-info component contains any additional information about the content. The proposed HBFR routing framework adaptively performs reactive and proactive content discovery based on content characteristics. It uses Bloom filters to announce the popular prefixes, and the announcement is restricted within a geographical region. The network is divided into hierarchical geographic regions to restrict the distribution of messages in order to reduce overhead. Vehicles in a single region form a cluster, and every vehicle in that region is a member of that cluster. Contents advertisement by a vehicle includes its region information, and it adds the content prefixes using Bloom filters in the content advertisements. These Bloom filter based advertisements are shared between regions to obtain a full view of the contents as well as their respective regions to easily discover and communicate the desired content.

The authors in [82] proposed the content-segmentation reassembly and reliable content delivery scheme by scheduling the retransmission of interests for lost-data messages. To quickly recover lost-data messages, the interests are retransmitted depending on the dynamic round-trip time (RTT) of the interest data exchange. The average RTT over a multi-hop path is estimated as a weighted moving average of RTT samples, and based on this information the retransmission time out is scheduled.

Last-encounter content routing (LER) in [83] is an opportunistic geo-inspired routing scheme for CCN-enabled VANET. Each vehicle that runs LER maintains two tables called the "content list" and the "last encounter list" (LEL). The content list contains the list of all contents that the vehicle holds and is shared within one-hop neighbors. The LEL enlists the content and information received from neighboring vehicles and enlists the provider ID, encounter position, and time. A content provided by multiple vehicles has multiple entries in the LEL. The interests are forwarded toward the location provided in the message.

A content-centric networking scheme for vehicular networks that uses provider information has been proposed in [84]. The scheme is proposed by the same authors and is an extension of the one proposed in [78]. This is the reason the scheme uses the same messages as in [78]: *A-Int*, *B-Int*, and *C-Obj*. The scheme uses provider selection mechanisms called "most responsive provider" (CCVN-MRP) and "nearest most responsive provider" (CCVN-NMRP). CCVN-MRP maintains the provider ID performed by the data message. The content-provider table is maintained at each node and stores content name, provider ID, and response counter.

The response counter is the number of times the provider replied to a data message for the said content. A consumer selects the provider with the highest response value, and that information of the selected provider is sent in the interest message. In contrast, the CCVN-NMRP scheme uses response count and hop-distance between the provider and consumer vehicles to select the potential provider. In CCVN-MRP, a consumer vehicle selects the nearest and most responsive provider.

The results of the vehicular NDN implementation in vehicles have been presented in [85]. The implementation uses various wireless faces, i.e., Wi-Fi, WiMAX, IEEE802.11p, etc., and the messages are transmitted over all the available faces to avoid disruptions caused by intermittent connectivity. The experiments have been performed in the vehicular network with no mobility, vehicles moving in a platoon, and vehicles moving around the University of California, Los Angeles (UCLA) campus.

In [86], the authors proposed a forwarding scheme that fetches data from a plethora of providers using digital map information. Navigo binds the NDN data names to the producers' geographic area(s). It uses a shortest-path algorithm to forward interests to the geographic area of the potential provider. The authors also claim the application of an adaptive best data provider discovery and selection scheme from multiple geographic areas.

The traffic violation ticketing (TVT) application for CCN-enabled vehicular networks has been proposed in [87]. It discusses the use of CCN interest and data messages used by police officers to issue violation tickets to drivers who commit violations. Additional data structures are maintained to achieve this application perspective. The same authors proposed and evaluated a hierarchal and hash-based content-naming scheme for vehicular networks [88]. The naming scheme encompasses the provider's identity, different components that represent the content attributes, and the spatio-temporal resolution of contents. A small hash component is also part of the name, which helps to precisely identify the content. Along with that, a compact tri-based name-management scheme has been adapted to manage the content name to perform speedy search, delete, and add name in the name prefix tables. Analysis shows that the proposed name-management scheme is more suitable for variable length name prefix management in content-centric networks.

A robust interest forwarder selection (RUFS) scheme for vehicular ad hoc networks has been proposed in [89]. RUFS mitigates the interest broadcast storm by selecting the suitable next-hop forwarder vehicle. In RUFS, each vehicle shares its satisfied-interests statistics with neighboring nodes. This information is managed in the neighbors-satisfied list (NSL). The authors also summarized the problems, challenges, and future perspectives of CCN and NDN in VANETs in [90]. ICN schemes for vehicular networks discussed previously are summarized in Table 3.2.

Table 3.2 CCN-/NDN-based schemes for vehicular networks

Name	ICN architecture	Scheme	Comparison	Network type	Simulator
NMND [72]	NDN	Collect vehicles' sensory information from the infrastructure (RSU)-supported vehicular network	MobileIP	Infrastructure-supported vehicular network	Qualnet
Hybrid VANET [74]	CCN	Implemented push-based event packet for CCN-based VANETs	–	VANETs	NS-3
Vehicular naming [75]	CCN/NDN	Hierarchical naming scheme for vehicular network representing temporal and spatial scope information	–	VANETs	–
CarSpeak [76]	MAC layer multiresolution naming	Collects sensory information from neighboring vehicles and nearby infrastructure	802.11 and 802.11 + naming	Golf Car and 10 iRobots connected with Xbox 360 Kinect sensors	Implemented on robots and Golf Car
CRoWn [77]	CCN	Forwarding strategy for V2V and V2I; it uses advertise, discover, and transfer content (*A-Int*, *B-Int*, and *C-Obj*) messages	Legacy TCP/IP-based architecture using AODV	Infrastructure-supported vehicular network	NS-2
CCVN [78]	CCN	Forwarding strategy with consumer- and provider-driven handover schemes; it uses the same messages as [vccn5]	Legacy TCP/IP-based architecture using AODV	Infrastructure-supported vehicular network	NS-2
CCVN [79]	CCN	Evaluates effectiveness of CCN in VANETs; it shows preliminary simulation results	Legacy TCP/IP-based architecture	Infrastructure-supported vehicular network	NS-2
V2V NDN [80]	NDN	Set of timers is used to avoid collision of NDN messages	–	V2V highway VANET scenario	NS-3 (ndnSIM)
HBFR [81]	CCN	Hierarchical Bloom-filter routing uses Bloom filters to advertise and communicated the contents in region-based clusters	CCN	VANET	Qualnet

(continued)

Table 3.2 (continued)

Name	ICN architecture	Scheme	Comparison	Network type	Simulator
CCN retransmission [82]	CCN	Content segmentation/reassembly and reliable content delivery (interest-retransmission scheme for lost-data messages)	–	Infrastructure-supported vehicular network	NS-2
LER [83]	CCN	Last-encounter content Routing is opportunistic geo-inspired content-based routing	Flooding	VANET	NS-3
CCVN-(MRP and NMRP) [84]	CCN	The provider is selected based on the large number of responses, number of responses, and the smallest hop distance	CCN	Infrastructure-supported vehicular network	NS-2
V-NDN [85]	NDN	NDN implementation over vehicular network	–	V2I, I2V, and Infrastructure-supported V2V	UCLA test bed
NAVIGO [86]	NDN	Geographic area based forwarding scheme for NDN-enabled VANETs	GPSR	Infrastructure-supported vehicular network	NS-3 (ndnSIM)
TVT [87]	CCN	Application of CCN in traffic violation ticketing use case	–	VANET	–
Hierarchical and hash-based Naming [88]	CCN	Hierarchical and hash-based naming scheme using compact Trie management scheme for vehicular CCN	Simple Trie and Bloom filters	VANET	C/C++
RUFS [89]	CCN	RobUst interest-forwarder selection scheme to mitigate interest-forwarder broadcast storm	CCN, DR-based, and NAIF	VANET	

References

1. CCNx—Content Centric Networking. url https://www.ccnx.org/
2. Jacobson V, Smetters DK, Thornton JD, Plass MF, Briggs NH, Braynard RL (2009) Networking named content. In: Proceedings of the 5th international conference on Emerging networking experiments and technologies (CoNEXT'09). ACM, New York, NY, USA, pp 1–12
3. Named Data Networking (NDN)—A future internet architecture. url http://named-data.net/, NSF's Future Internet Architecture Program
4. Content-Centric Networking CCNx Reference Implementation. url https://github.com/ProjectCCNx/ccnx
5. Kim K, Choi S, Kim S, Roh B-h (2013) A push-enabling scheme for live streaming system in content-centric networking. In: Proceedings of the 2013 workshop on Student workshop (CoNEXT Student Workshop'13). ACM, New York, NY, USA, pp 49–52
6. Francois J, Cholez T, Engel T (2013) CCN traffic optimization for IoT. In: 2013 Fourth international conference on the network of the future (NOF), pp 1–5, 23–25 Oct 2013
7. Teubler T, Hail MAM, Hellbrück H (2013) Efficient data aggregation with CCNx in wireless sensor networks. In: Advances in communication networking, vol 8115. Lecture notes in computer science, pp 209–220
8. Mosko M, Solis I, Uzun E, Wood C (2015) CCNx 1.0 protocol architecture. Technical report, Aug 2015. url http://www.ccnx.org/pubs/CCNxProtocolArchitecture.pdf
9. Zhang B, Afanasyev A, Burke J, Jacobson V, Crowley P, Papadopoulos C, Wang L, Zhang B (2010) Named data networking
10. Ren J et al (2014) MAGIC: a distributed MAx-Gain In-network caching strategy in information-centric networks. IEEE INFOCOM NOM workshop
11. Bernardini C, Silverston T, Festor O (2014) Socially-aware caching strategy for content centric networking. IFIP networking
12. Bernardini C, Silverston T, Festor O (2013) MPC: popularity-based caching strategy for content centric networks. IEEE ICC
13. Psaras I, Chai WK, Pavlou G (2012) Probabilistic in-network caching for information-centric networks. In: Proceedings of the second edition of the ICN workshop on Information-centric networking. ACM, New York
14. Chai WK et al (2012) Cache less for more in information-centric networks. In: NETWORKING 2012. Springer, Berlin, pp 27–40
15. Private definitions for ccnd—the CCNx daemon "ccnd_private.h". url https://github.com/ProjectCCNx/ccnx/blob/master/csrc/ccnd/ccnd_private.h
16. Rowley Jennifer (2007) The wisdom hierarchy: representations of the DIKW hierarchy. J Inf Sci 33(2):163–180. doi:10.1177/0165551506070706
17. Mosko M, Solis I, Uzun E (2015) CCN 1.0 protocol architecture. Palo Alto Research Center. url http://ccnx.org/pubs/ICN_CCN_Protocols.pdf
18. Mosko M (2015) CCNx content object chunking. ICNRG, Internet-Draft, draft-mosko-icnrg-ccnxchunking-01, 1 July 2015
19. Mosko M, Scott G, Solis I, Wood C (2015) CCNx manifest specification. ICNRG, Internet-Draft, draft-wood-icnrg-ccnxmanifests-00, 24 June 2015. url http://www.ccnx.org/pubs/draft-wood-icnrg-ccnxmanifests-00.html
20. Mosko M, Solis I, Mahadevan P, Uzun E (2014) CCNx 1.0 naming: transforming network addresses to application value. Technical report, PARC, 16 March 2014
21. Zhang H, Quan W, Guan J, Xu C, Song F (2016) Uniform information with a hybrid naming (hn) scheme. (https://tools.ietf.org/html/draft-zhang-icnrg-hn-01) draft-zhang-icnrg-hn-01.txt, Expires: 7 April 2016
22. Ding S, Chen Z, Liu Z (2012) Parallelizing FIB lookup in content centric networking. 2012 Third international conference on networking and distributed computing (ICNDC). IEEE, pp 6–10

23. Mosko M (2015) CCNx semantics. IETF Internet-Draft, ccnx-mosko-semantics-01, 21 Jan 2015
24. Mosko M (2015) TLV packet format. IETF Internet-Draft, ccnx-mosko-tlvpackets-01, 21 Jan 2015
25. Mosko M (2015) Labeled content information. ICNRG, Internet-Draft, draft-mosko-icnrg-ccnxlabeledcontent-00, 13 July 2015
26. Abu AJ, Bensaou B, Wang JM (2014) Interest packets retransmission in lossy CCN networks and its impact on network performance. ICN'14, 24–26 Sept 2014, Paris, France, pp 167–176
27. Bernardini C, Silverston T, Festor O (2013) Cache management strategy for CCN based on content popularity. In: Emerging management mechanisms for the future internet, vol 7943. Lecture notes in computer science. Springer, Berlin, pp 92–95
28. Miorandi D, Sicari S, De Pellegrini F, Chlamtac I (2012) Internet of things: vision, applications and research challenges. Ad hoc Networks 10(7):1497–1516, Sept 2012. ISSN 1570-8705, http://dx.doi.org/10.1016/j.adhoc.2012.02.016
29. IERC-European Research Cluster on the Internet of Things (2011) Internet of things—Pan European Research and Innovation Vision, 2011. url http://www.internet-of-things-research.eu/pdf/IERC_IoT-Pan%20European%20Research%20and%20Innovation%20Vision_2011_web.pdf
30. Wilson S (2013) Rising tide—Exploring pathways to growth in the mobile semiconductor industry, 6 Nov 2013. url http://dupress.com/articles/rising-tide-exploring-pathways-to-growth-in-the-mobile-semiconductor-industry/
31. Waltari OK (2013) Content-centric networking in the internet of things. MSc thesis, Department of Computer Science, University of Helsinki, 25 Nov 2013. url http://hdl.handle.net/10138/42303
32. Quevedo J, Corujo D, Aguiar R (2014) A case for ICN usage in IoT environments. In: Global Communications Conference (GLOBECOM), 2014. IEEE, 8–12 Dec 2014, pp 2770–2775
33. Baccelli E, Mehlis C, Hahm O, Schmidt TC, Wählisch M (2014) Information centric networking in the IoT: experiments with NDN in the wild. In: Proceedings of the 1st international conference on Information-centric networking (ICN'14). ACM, New York, NY, USA, pp 77–86
34. CCN Lite: lightweight implementation of the content centric networking protocol, 2014. url http://ccn-lite.net
35. RIOT: the friendly operating system for the internet of things. url http://www.riot-os.org/
36. Amadeo M, Campolo C, Molinaro A (2014) Multi-source data retrieval in IoT via named data networking. In: Proceedings of the 1st international conference on Information-centric networking (ICN'14). ACM, New York, NY, USA, pp 67–76
37. Katsaros K, Chai W, Wang N, Pavlou G, Bontius H, Paolone M (2014) Information-centric networking for machine-to-machine data delivery: a case study in smart grid applications. In: Network, vol 28, no 3. IEEE, May–June 2014, pp 58–64
38. Yu K, Zhu L, Wen Z, Mohammad A, Zhou Z, Sato T (2014) CCN-AMI: performance evaluation of content-centric networking approach for advanced metering infrastructure in smart grid. In: 2014 IEEE international workshop on applied measurements for power systems proceedings (AMPS), 24–26 Sept 2014, pp 1–6
39. Yu K, Arifuzzaman M, Wen Z, Zhang D, Sato T (2015) A key management scheme for secure communications of information centric advanced metering infrastructure in smart grid. In: IEEE transactions on instrumentation and measurement, vol 64, no 8, Aug 2015, pp 2072–2085
40. Chai WK, Katsaros KV, Strobbe M, Romano P, Ge C, Develder C, Pavlou G, Wang N (2015) Enabling smart grid applications with ICN. 2nd ACM conference on information-centric networking (ICN 2015), 30 Sept–2 Oct 2015, pp 207–208
41. Rault T, Bouabdallah A, Challal Y (2014) Energy efficiency in wireless sensor networks: a top-down survey. Comput Netw 67(4):104–122

42. Kafi MA, Djenouri D, Ben-Othman J, Badache N (2014) Congestion control protocols in wireless sensor networks: a survey. In: Communications surveys & tutorials, vol 16, no 3. IEEE, pp 1369–1390, Third Quarter 2014
43. Rawat P, Singh KD, Chaouchi H, Bonnin JM (2014) Wireless sensor networks: a survey on recent developments and potential synergies. J Supercomput 68(1):1–48, 09 Oct 2013
44. Singh SP, Sharma SC (2015) A survey on cluster based routing protocols in wireless sensor networks. In: Procedia computer science, vol 45, pp 687–695
45. Butun I, Morgera SD, Sankar R (2014) A survey of intrusion detection systems in wireless sensor networks. In: Communications surveys & tutorials, vol 16, no 1. IEEE, pp 266–282, First Quarter 2014
46. Saadallah B, Lahmadi A, Festor O (2012) CCNx for Contiki: implementation details. Technical report RT-0432, INRIA, p 52
47. Ren Z, Hail MA, Hellbruck H (2013) CCN-WSN—A lightweight, flexible Content-Centric Networking protocol for wireless sensor networks. In: 2013 IEEE eighth international conference on intelligent sensors, sensor networks and information processing, 2–5 April 2013, pp 123–128
48. Meijers JP, Amadeo M, Campolo C, Molinaro A, Paratore SY, Ruggeri G, Booysen MJ (2013) A two-tier content-centric architecture for wireless sensor networks. In: 2013 21st IEEE international conference on network protocols (ICNP), 7–10 Oct 2013, pp 1–2
49. Amadeo M, Campolo C, Molinaro A, Mitton N (2013) Named data networking: a natural design for data collection in wireless sensor networks. In: Wireless Days (WD), 2013 IFIP, 13–15 Nov 2013, pp 1–6
50. Abidy Y, Saadallahy B, Lahmadi A, Festor O (2014) Named data aggregation in wireless sensor networks. In: Network operations and management symposium (NOMS), 2014 IEEE, 5–9 May 2014, pp 1–8
51. Burke J, Gasti P, Nathan N, Tsudik G (2014) Secure sensing over named data networking. In: Proceedings of the 13th IEEE international symposium on network computing and applications (NCA)
52. Baumgartner T, Chatzigiannakis I, Fekete S, Koninis C, Kröller A, Pyrgelis A (2010) Wiselib: a generic algorithm library for heterogeneous sensor networks. In: Sá Silva J, Krishnamachari B, Boavida F (eds) Proceedings of the 7th European conference on wireless sensor networks (EWSN'10). Springer, Berlin, pp 162–177
53. Contiki: The open source OS for the internet of things. url http://www.contiki-os.org/
54. Dorronsoro B, Ruiz P, Danoy G, Pigne Y, Bouvry P (2014) Evolutionary algorithms for mobile ad hoc networks. Wiley, New York
55. Mahmood BA, Manivannan D (2015) Position based and hybrid routing protocols for mobile ad hoc networks: a survey. Wireless Personal Commun 83(2):1009–1033, 21 Feb 2015
56. Reina DG, Askalani M, Toral SL, Barrero F, Asimakopoulou E, Bessis N (2015) A survey on Multihop ad hoc networks for disaster response scenarios. Int J Distrib Sensor Netw 2015:16 p
57. Attia R, Rizk R, Ali HA (2015) Internet connectivity for mobile ad hoc network: a survey based study. In: Wireless networks, vol 21, No 7, 1 Oct 2015, pp 2369–2394
58. Ruiz P, Bouvry P (2015) Survey on broadcast algorithms for mobile ad hoc networks. ACM Comput Surv 48(1):35 p, Article 8
59. Meisel M, Pappas V, Zhang L (2010) Ad hoc networking via named data. In: Proceedings of the fifth ACM international workshop on mobility in the evolving internet architecture (MobiArch'10). ACM, New York, NY, USA, pp 3–8
60. Oh SY, Lau D, Gerla M (2010) Content centric networking in tactical and emergency MANETs. In: IFIP wireless days (WD), 20–22 Oct 2010, pp 1–5
61. Varvello M, Rimac I, Lee U, Greenwald L, Hilt V (2011) On the design of content-centric MANETs. In: 2011 Eighth international conference on wireless on-demand network systems and services (WONS), 26–28 Jan 2011, pp 1–8
62. Amadeo M, Molinaro A (2011) CHANET: a content-centric architecture for IEEE 802.11 MANETs. In: 2011 International Conference on the network of the future (NOF), 28–30 Nov 2011, pp 122–127

63. Cianci I, Grieco LA, Boggia G (2012) CCN—Java opensource kit EmulatoR for wireless ad hoc networks. In: Proceedings of the 7th international conference on future internet technologies (CFI'12). ACM, New York, NY, USA, pp 7–12

64. Adem O, Kang S-j, Ko Y-B (2013) Packet loss avoidance in content centric mobile adhoc networks. In: Proceedings of the 15th international conference on advanced communication technology (ICACT), 2013, 27–30 Jan 2013, pp 245–250

65. Yu Y-T, Dilmaghani RB, Calo S, Sanadidi MY, Gerla M (2013) Interest propagation in named data manets. In: International conference on computing, networking and communications (ICNC), 2013, 28–31 Jan 2013, pp 1118–1122

66. Amadeo M, Molinaro A, Ruggeri G (2013) E-CHANET: routing, forwarding and transport in information-centric multihop wireless networks. In: Computer communications, vol 36, No 7, pp 792–803, April 2013

67. Varvello M, Schurgot M, Esteban J, Greenwald L, Guo Y, Smith M, Stott D, Wang L (2013) SCALE: a content-centric MANET. In: 2013 IEEE conference on computer communications workshops (INFOCOM WKSHPS), 14–19 April 2013, pp 29–30

68. Kim J, Shin D, Ko Y-B (2013) TOP-CCN: topology aware content centric networking for mobile ad hoc networks. In: 2013 19th IEEE international conference on networks (ICON), 11–13 Dec 2013, pp 1–6

69. Detti A, Tassetto D, Melazzi NB, Fedi F (2015) Exploiting content centric networking to develop topic-based, publish–subscribe MANET systems. In: Ad hoc networks, vol 24, Part B, Jan 2015, pp 115–133

70. Amadeo M, Campolo C, Molinaro A (2015) Forwarding strategies in named data wireless ad hoc networks: design and evaluation. J Network Comput Appl 50:148–158

71. IEEE Standard for Wireless Access in Vehicular Environments (WAVE)—Networking services—Redline. In: IEEE Std 1609.3-2010 (Revision of IEEE Std 1609.3-2007)—Redline, 30 Dec 2010, pp 1–212

72. TalebiFard P, Leung VCM, Amadeo M, Campolo C, Molinaro A (2015) Information-centric networking for VANETs, Chap 17. In: Campolo C, Molinaro A, Scopigno R (eds) Vehicular ad hoc networks. Springer, Berlin, pp 503–524

73. Wang J, Wakikawa R, Zhang, L (2010) DMND: collecting data from mobiles using named data. In: IEEE vehicular networking conference (VNC), 2010, 13–15 Dec 2010, pp 49–56

74. Arnould G, Khadraoui D, Habbas Z (2011) A self-organizing content centric network model for hybrid vehicular ad-hoc networks. In: Proceedings of the first ACM international symposium on design and analysis of intelligent vehicular networks and applications (DIVANet'11). ACM, New York, NY, USA, pp 15–22

75. Wang L, Wakikawa R, Kuntz R, Vuyyuru R, Zhang L (2012) Data naming in vehicle-to-vehicle communications. In: IEEE conference on computer communications workshops (INFOCOM WKSHPS), 25–30 March 2012, pp 328–333

76. Kumar S, Shi L, Ahmed N, Gil S, Katabi D, Rus D (2012) CarSpeak: a content-centric network for autonomous driving. SIGCOMM Comput Commun Rev 42(4):259–270

77. Amadeo M, Campolo C, Molinaro A (2012) CRoWN: content-centric networking in vehicular ad hoc networks. IEEE Commun Lett 16(9):1380–1383

78. Amadeo M, Campolo C, Molinaro A (2012) Content-centric vehicular networking: an evaluation study. In: Third international conference on the network of the future (NOF), 21–23 Nov 2012, pp 1–5

79. Amadeo M, Campolo C, Molinaro A (2012) Content-centric networking: is that a solution for upcoming vehicular networks? In: Proceedings of the ninth ACM international workshop on vehicular inter-networking, systems, and applications (VANET'12). ACM, New York, NY, USA, pp 99–102

80. Wang L, Afanasyev A, Kuntz R, Vuyyuru R, Wakikawa R, Zhang L (2012) Rapid traffic information dissemination using named data. In: Proceedings of the 1st ACM workshop on emerging name-oriented mobile networking design—Architecture, algorithms, and applications (NoM'12). ACM, New York, NY, USA, pp 7–12

81. Yu Y-T, Li X, Gerla M, Sanadidi MY (2013) Scalable VANET content routing using hierarchical bloom filters. In: 9th international wireless communications and mobile computing conference (IWCMC), 1–5 July 2013, pp 1629–1634
82. Amadeo M, Campolo C, Molinaro A (2013) Design and analysis of a transport-level solution for content-centric VANETs. In: Proceedings of the IEEE international conference on communications workshops (ICC), 9–13 June 2013, pp 532–537
83. Yu Y-T, Li Y, Ma X, Shang W, Sanadidi MY, Gerla M (2013) Scalable opportunistic VANET content routing with encounter information. In: 2013 21st IEEE international conference on network protocols (ICNP), 7–10 Oct 2013, pp 1–6
84. Amadeo M, Campolo C, Molinaro A (2013) Enhancing content-centric networking for vehicular environments. Comput Networks 57(16):3222–3234, 13 Nov 2013
85. Grassi G, Pesavento D, Pau G, Vuyyuru R, Wakikawa R, Zhang L (2014) VANET via named data networking. In: IEEE conference on computer communications workshops (INFOCOM WKSHPS), 27 April–2 May 2014, pp 410–415
86. Grassi G, Pesavento D, Pau G, Zhang L, Fdida S (2015) Navigo: interest forwarding by geolocations in vehicular named data networking. IEEE 16th international symposium on "a world of wireless, mobile and multimedia networks" (WoWMoM), June 2015, pp 1–10
87. Ahmed SH, Yaqub MA, Bouk SH, Kim D (2015) Towards content-centric traffic ticketing in VANETs: an application perspective. In: 2015 Seventh international conference on ubiquitous and future networks (ICUFN), 7–10 July 2015, pp 237–239
88. Bouk SH, Ahmed SH, Kim D (2015) Hierarchical and hash based naming with Compact Trie name management scheme for vehicular content centric networks. Comput Commun. Available online, 3 Oct 2015
89. Ahmed SH, Bouk SH, Kim Dongkyun (2015) RUFS: RobUst forwarder selection in vehicular content-centric networks. IEEE Commun Lett 19(9):1616–1619
90. Bouk SH, Ahmed SH, Kim D (2015) Vehicular content centric network (VCCN): a survey and research challenges. In: Proceedings of the 30th annual ACM symposium on applied computing (SAC'15). ACM, New York, NY, USA, pp 695–700

Chapter 4
Future Aspects

Syed Hassan Ahmed, Safdar Hussain Bouk and Dongkyun Kim

Abstract Due to the early stage in development, content-centric networking (CCN) and its variants e.g., named-data network (NDN), are undergoing rigorous modifications to make them applicable to various future networks. In the previous chapter, we discussed the basic working principle of CCN as well as different CCN-based solutions for diverse application scenarios. In this chapter, we identify future research aspects and the issues that have been partially addressed or have not been properly tackled by researchers. In addition, we also provide a road map for researchers of the relevant field.

Keywords Content · Naming · Forwarding · CCN data structures · Research issues

4.1 Research Directions

There are different categories of future aspects in content-centric networking (CCN) that must be addressed [1]; however, we divide them per their related CCN components including: content (e.g., chunking, discovery, manifest, multi-homing, etc.); naming (e.g., hierarchical, hash-based, attribute-based, name-management scheme, etc.); data structures (e.g., PIT management, FIB management, etc.); CCN message forwarding; content caching; dynamic network topology (e.g. provider mobility, consumer mobility, effects of dynamic topology on content forwarding, etc.); and security and privacy. Here we briefly discuss the potential future research directions of CCN; however, readers can refer to [1–3] for further details. CCN research directions are briefly discussed in this section.

(a) *Content*

Content can be any binary object or binary stream such as any type of file, audio stream, video stream, etc. In general, most of the research is focused on how to name a content and efficiently forward the content in CCN. There are some basics about content that should be addressed, e.g., chunking, manifest, multi-homing, etc. Content chunking divides a large content into multiple small identifiable pieces to

© The Author(s) 2016
S.H. Ahmed et al., *Content-Centric Networks*, SpringerBriefs in Electrical and Computer Engineering, DOI 10.1007/978-981-10-0066-9_4

efficiently forward the content from the consumer to the provider. It also provides a means to evenly disperse the content in the distributed network caches. There is no CCN standard to divide a content into the chunks of universally equal size. Some of the research works assume that a content chunk is equal to the size of an MTU. The fact is that the MTU size may vary based on the face type. Therefore, if a node receives multiple chunks of content that do not fit the outgoing face, the chunks require rearrangement. It also poses additional issues that must be resolved for the CCN transport layer. These issues are enlisted, but not limited to, as follows:

- Synchronization between different chunk providers or sources
- Receiver-driven synchronization
- Cache-management policies for chunks
- Chunk-level security checks
- Chunk-based or content-based identification
- Content or chunk-based authentication
- Content manifest, etc.

The above-mentioned issues have influenced and pose complexity to many other aspects that affect efficient content communication in CCN.

(b) *Naming*

Content-object retrieval in CCN usually involves two stages: discovery and delivery of the content using interest and data messages. The content is discovered through its name, which uniquely identifies it in the network. This is the reason that a content name is a mandatory TLV in the interest and data messages. An interest message with a content name is used to discover content. Content delivery involves rules to route contents on the network, and it also uses the content name to make routing decisions.

There are different ways to represent the name including hierarchical, flat, and attribute-based. The authors in [4] discuss the pros and cons, management, and respective issues related to each of these schemes. However, here we focus on future directions concerning the hierarchical naming scheme used by the CCN architecture.

Due to the hierarchical nature of the naming scheme used in CCN, names can easily be aggregated and naturally have the longest prefix matching feature. Along with these inherent features, there are still some issues that must be resolved. Following are some of the future research directions that must be addressed:

- Evaluation to measure the effectiveness and scalability to the Internet level is still required for hierarchical names.
- The security information is not part of the CCN naming because it assumes explicit security provided by the content itself. Therefore, hybrid schemes are required for CCN naming to provide security information within the name.
- Longest prefix matching on the name-string is time consuming and requires feasible solutions to minimize the search time.

- Efficient data structures are required to optimize the memory use for hierarchical names with varying prefix sizes.
- Related to the previous point, the data structure should also be effective enough to perform speedy management of hierarchical names, such as adding and deleting name prefixes, because the arrival rate of the names can be much higher in the global-scale Internet.
- Number, position, and size of the hierarchical name components are not fixed, and inferring the name components is still application-dependent. Hence, there is a need to standardize naming structure and inferencing rules.
- There are some applications, e.g., vehicular networks, WSNs, etc., where spatial and temporal scope contents are required. A few schemes discuss these issues; however, more research in this aspect is required.
- Hardware-level implementation and management of hierarchical names is still a pending issue.

(c) *Data Structures*

Name prefixes are managed in the PIT and FIB to forward interest and data messages in the network. named-data network (NDN), the variant of CCN, keeps track of many parameters, e.g., NONCE list, dead NONCE list, timers, etc., to avoid routing loops [5, 6]. The incoming name plus face information and outgoing face plus prefix information are stored the PIT and FIB tables of CCN. These tables are refreshed, and stale information is removed. The size of these tables and the data-delivery rate are directly related to the refresh duration. Therefore, this effect must be tested in varying network scenarios, e.g., wired, wireless, dynamic topology networks, etc. The size and look-up efficiency of these data structures is still an open issue in CCN.

Stateful forwarding in CCN requires some information in these tables and has been investigated in [7, 8]. Therefore, the effect and efficiency of the state information on interest and data forwarding delay and the size of these tables should be investigated.

(d) *CCN Message Forwarding*

CCN uses the FIB and PIT to forward interest and data messages in the network. Basically, most of the interest messages are forwarded through a longest name prefix matching process within the FIB. The interest message is forwarded to the face associated with the matching FIB entry. In case of multiple outgoing faces, the forwarding strategy selects the most stable face (note that the description *of* "most suitable" depends on which parameter is being used to rank the outgoing face as most suitable, e.g., interest-satisfaction rate, most recent activity over face, etc.) outgoing face. However, there may be a possibility that a subset of the faces may be selected to forward the interest message. Therefore, an efficient forwarding mechanism is required that selects the outgoing face not only on FIB match but that also considers network and neighborhood parameters.

Moreover, the interest is forwarded between the consumer and the provider node through multiple paths, and the consumer may receive multiple copies of the interest message at varying time instances. Normally, a provider or a caching node in the CCN replies with data to the first interest message that it receives, and the following received copies of the interest message are dropped by the provider node. In this case, there may a possibility that the downstream direction in that path may not be suitable to forward data message. Therefore, a provider node must hold the data message and select the best path based on the path information received through multiple copies of the interest message from different paths. The duration to hold the data message and selection of the suitable path(s) is still an open issue as well as a possible future research direction.

Interest overhead is one of the issues that should be mitigated in CCN. An interest message is forwarded by each node that receives the interest message and has no PIT entry. This may lead to a very high interest overhead, which can be minimized by limiting the number of interest forwarders to be investigated. In addition, the new transport modes—including anycast (any-to-any), multicast (i.e., one-to-many), and concast (i.e., many-to-one)—should be examined [9]. Preliminary investigations into these topics are presented in [10–12], but more investigations are required in this regard. Future investigations could also study these delivery modes and their effect on the congestion control, flow control, etc.

Likewise, distributed caching in CCN may also pose a situation where multiple providers for a single interest is possible. In this case, the consumer must select the most suitable provider among the subset. When an interest is issued for a content object that can be satisfied by the subset of providers, then synchronization between that subset of providers is necessary to increase content-transfer efficiency.

(e) *Content Discovery*

CCN uses the content name in the interest message to discover the content in the network. As previously discussed, interest messages are routed based on the longest prefix matching process in content name tables, i.e., the PIT and FIB, which are maintained at each CCN-enabled node in the network. The provider node, or any intermediate node that has a matching copy of the content in its cache, sends the data message to the consumer node following the reverse path. No location information is available for the desired content provider in the network. One solution to cope this problem is to maintain additional information regarding content providers in the data structure(s) based on previously satisfied interests. The next interests demanding the same contents are forwarded toward those providers [13]. However, if a node has no provider information available, then it follows the conventional CCN-forwarding mechanism [14].

Content discovery becomes more challenging and may be mandatory in the distributed network where contents are generated dynamically. The authors in [15] proposed enumeration request discovery (ERD) and regular interest discovery (RID) schemes. In ERD, the consumer sends an interest message requesting enumeration of the first-level prefix in the content name and requests for the next-level components available in local or remote repositories. The whole name can be discovered iteratively starting from the top level. After receiving the next-level name components, the consumer can include or exclude the next-level components in the following interest. In this manner, the consumer locates the cache that has the specific component, or no more information is received in response to the interest message. In contrast, in RID the consumer sends regular interest messages, and it receives the initial content segment in response. Then the consumer sends interest messages either including or excluding the additional prefix components in the content name to request the additional content. The authors in [cd02] proposed a content-discovery scheme, called SCAN, where each node exchanges the content store information within the neighborhood. This content store of information is organized in the data structure, "called the cache information base" (CIB), in addition to the FIB, PIT, and CS. Bloom filters are used to minimize the information size in the advertisement messages. A variant of SCAN has also been proposed in the same work, called "SCAN-ch," to discover and store chunk-level content information.

The above-mentioned schemes provide the base to explore a topic to answer the following questions.

- How can contents can be discovered with minimum overhead?
- How can content-discovery delay be reduced?
- What are the efficient methods to manage neighborhood- or network-wide content information on a node?
- What is the effect of content discovery on content and communication efficiency in CCN?
- Is content discovery feasible in highly mobile or highly dynamic topology networks, etc.?

(f)   *Dynamic Network Topology*

Dynamic network topology is defined as a network topology that varies over time due to node mobility, node failures, link failures, etc. Much research has been performed to handle network dynamicity in the past, and many solutions have been proposed for wired and wireless networks. There are few initial proposals that tackle mobility in CCN, for example, [16–21]. It is argued that consumer mobility is inherently handled by most CN architectures (especially CCN/NDN) because they use a pull-based or consumer-driven scheme. Previously cited work mostly focuses on consumer and/or provider mobility where they try to achieve higher interest satisfaction and delivery rate and minimize latency. It is claimed in CCN architecture documentation that CCN inherently handles mobility by detaching the location information binding with the content. However, the receiver's and provider's mobility may have a high effect on interest- and data-message transmission. Another side effect of dynamic topology, other than data/interest message loss, is that it leaves traces of routing information in the tables at intermediate nodes, which may increase communication delay due to high look-up cost. The issues related to provider mobility that have been identified and need to be addressed are (1) locating the provider node all of the time and (2) maintaining the connectivity until the complete content is received. In the absence of position information and route updates in CCN, CNN requires new mechanism(s) to address these issues. Another question waiting for an answer is that of the examination and benefits of caching in dynamic topology networks. In addition, the new schemes are required to ensure the intactness of content in the presence of mobility.

(g) *Content Caching in CCN*

Content caching is not a new topic in communications networks and has been intensively investigated in the past. Most of the studies are focused on caching in web applications and peer-to-peer networks. Different caching schemes—e.g., least frequently/recently used (LFU/LRU), most frequently/recently used (MFU/MRU), leave copy down (LCD), popularity-based caching, etc.—have been investigated in CCN [22]. It has been argued and serious concerns discussed in [23] that extensive use of caching in CCN will not confer considerable benefits. Therefore, more investigations are required to explore caching based on various communication patterns in CCN. Caching may pose more challenges regarding efficient cache-size use when the network topology is unpredictable, and there is a need for further research on these issues. Information popularity is another issue that must be investigated regarding how dynamic content popularity is decided. In the presence of different traffic patterns, an in-network cache is more challenging, and a few solutions are presented in [24–26].

To summarize, solicitation of in-network caching and replication schemes for CCN require new paradigms that jointly investigate routing, forwarding, and cache-management optimization i.e., the effect of cache locations on routing decisions, cache contention for varying information nature, etc.

(h) *Security and Privacy*

CCN claims to secure the content rather than the connection. In other words, instead of securing the connection between sender and receiver, CCN inherently secures the content by sending security-related information along with the content itself. Now the question is this: What is the content security? (or what security aspects are related to the content in CCN?).

CCN architecture requires public-key cryptography (PKC) to bind a public key with a content name. Public key cryptography is also used by most of the ICN architectures to provide security and privacy features. The main focus of the CCN is to secure the content object (either as whole content objects or as individual chunks of the content object). However, this poses the question of the necessity to digitally sign and encrypt every content as well as how to decide which content should be encrypted or signed? To secure the content, a producer or generator of the content digitally signs the content, and the publisher's information is also provided within the content. This avoids third-party dependency. The consumer uses this supplementary information with the content object to detect the content integrity—i.e., whether the content object was compromised by any node during the communication between producer and consumer—and to assure that the content is produced by a trusted publisher. The hierarchical names are human readable and can easily identify the publisher of the content. This can easily bind the name with the publisher. However, there must be some mechanism to verify whether or not the key associated with the name really belong to the publisher.

Because most ICN architectures, including CCN, rely on public key cryptography to provide security, then the paramount question is this: Who will be responsible for creating, distributing, and revoking these keys? In addition, if CCN relies on the trusted entities for name verification, then key management becomes a prime issue in CCN [27].

Another unexplored topic in the area of security is access control that specifies which node can access what cached information. The authors in [28] gave a detailed explanation of the access control in CCNx 1.0, called "modified hierarchical access control," which uses access control lists (ACLs) at nodes for the contents or segmented contents. However, it is required to have some more policies and/or schemes to specifically define access control in CCN [29], i.e., granting and revocation of access rights, managing rights, and resolving access-right conflicts, etc.

In mobile networks, especially ad hoc nature networks, security and the resolution of several security attacks is still a challenging issue. CCN claims and focuses on securing the content, rather than the connection, using public key encryption, which means that it promises that the content is the same as it asserts. This may raise more privacy issues because the content name is advertised in the interest messages. In infrastructure-supported wireless networks, privacy and trust issues are alleviated due to various points of management in the network, e.g., access routers, etc. However, in case of pure ad hoc and mobile networks without any infrastructure support require robust solutions to handle privacy and security threats. These solutions can be more challenging in constrained-device scenarios, e.g., limited processing power, small memory, limited bandwidth, etc.

Several security attacks have been identified for CCN/NDN, and they require robust solutions to prevent them, e.g., denial-of-service (DOS), distributed DOS (DDoS), sniffing and Watchlist, black hole, flooding, content pollution, etc. The readers are suggested to refer [30–33] to obtain security insight related to CCN.

(i) *Evaluation Methods*

Currently many researchers are pursuing and publishing solutions for CCN, and they evaluate their schemes through simulations, theoretical, and empirical evaluations. Only few schemes have been empirically evaluated due to time, budget, access, and other limitations. However, most of the schemes have been proposed for a varying range of network scenarios (IoT, VANET, WSN, MANET, etc.) and simulated in freely available or open-source simulators. The authors in [34] summarized ICN evaluation methodologies and surveyed different evaluation methods, tools, topology-selection matrices, etc.

Most of the solutions are evaluated through simulations and along with that, to keep fairness in evaluation, standard or baseline network scenarios should be considered in the simulations. The baseline scenarios for different network types and their scenario code is discussed in [35]. In addition to that, traffic load, content popularity, and different other metrics are also explored by the authors. To maintain fairness in simulation evaluation for the proposed scheme, it is necessary to have baseline scenarios and standard parameters. additional challenges in CNN may need to be explored because active, ongoing research on the topic is being pursued around the globe, which may uncover other needs.

4.2   Conclusion

In this book, we first provided an overview of Internet communications and its varying and emerging architectures. In Chap. 2, we briefly discussed different future Internet architectures and provided introductory details. Later we provided detailed aspects of the selected ICN architecture known as CCN. We gave a deep overview of its operations as well as all of the definitions in the third chapter. To be precise, we call Chapter the backbone chapter of the book. Finally, in this chapter, we discussed the existing issues to be considered as research challenges in the promising field of CCN. After reading this book, we expect that graduate students, professionals, and beginners will be able to extract the issues that require significant attention from academic and industrial researchers. We conclude this book by stating that upcoming research results in CCN should enhance the effectiveness, efficiency, and security of content communications.

References

1. Kutscher D, Eum S, Pentikousis K, Psaras I, Corujo D, Saucez D, Schmidt T, Waehlisch M (2016) ICN research challenges, draft-irtf-icnrg-challenges-02, ICNRG, Internet-Draft, Expires, 5 Mar 2016
2. Xylomenos G, Ververidis CN, Siris VA, Fotiou N, Tsilopoulos C, Vasilakos X, Katsaros KV, Polyzos GC (2014) A survey of information-centric networking research. Commun Surv Tutorials 16(2):1024–1049
3. Vasilakos AV, Li Z, Simon G, You W (2015) Information centric network: research challenges and opportunities. J Netw Comput Appl 52:1–10
4. Zhang H, Quan W, Guan J, Xu C, Song F (2016) Uniform information with a hybrid naming (hn) scheme, draft-zhang-icnrg-hn-03.txt, ICNRG Internet Draft, Expires, 7 Apr 2016
5. Yi C, Abraham J, Afanasyev A, Wang L, Zhang B, Zhang L (2014) On the role of routing in named data networking. In: Proceedings of the 1st international conference on information-centric networking (ICN'14). ACM, New York, NY, USA, pp 27–36
6. Garcia-Luna-Aceves JJ, Mirzazad-Barijough M (2015) Enabling correct interest forwarding and retransmissions in a content centric network. In: Proceedings of the eleventh ACM/IEEE symposium on architectures for networking and communications systems (ANCS'15). IEEE Computer Society, Washington, DC, USA, pp 135–146
7. Yi C, Afanasyev A, Moiseenko I, Wang L, Zhang B, Zhang L (2012) A case for stateful forwarding plane. NDN Technical Report NDN-0002, July 2012

8. Tsilopoulos C, Xylomenos G, Thomas Y (2014) Reducing forwarding state in content-centric networks with semi-stateless forwarding. In: Proceedings of IEEE INFOCOM, 2014, pp 2067–2075, 27 Apr–2 May 2014

9. Schmidt T, Waehlisch M (2012) Why we shouldn't forget multicast in name-oriented publish/subscribe, arXiv preprint arXiv: abs/1201.0349, URL: http://arxiv.org/abs/1201.0349, 10 Oct 2012

10. Tsilopoulos C, Gasparis I, Xylomenos G, Polyzos GC (2013) Efficient real-time information delivery in future internet publish-subscribe networks. In: International conference on computing, networking and communications (ICNC), 2013, pp 856–860, 28–31 Jan 2013

11. Garcia-Luna-Aceves JJ (2015) Efficient multi-source multicasting in information centric networks. In: 12th annual IEEE consumer communications and networking conference (CCNC), 2015, pp 245–249, 9–12 Jan 2015

12. Thomas Y, Tsilopoulos C, Xylomenos G, Polyzos GC (2013) Multisource and multipath file transfers through publish-subscribe internetworking. In: Proceedings of the 3rd ACM SIGCOMM workshop on information-centric networking (ICN'13). ACM, New York, NY, USA, pp 43–44

13. Lee M, Song J, Cho K, Pack S, Kwon TT, Kangasharju J, Choi Y (2015) Content discovery for information-centric networking. Comput Netw 83:1–14, 4 June 2015

14. Scott G (2014) CCNx 1.0 simple service discovery. Technical Report, Computing Science Laboratory, Palo Alto Research Center, 27 Feb 2014

15. Anastasiades C, Uruqi A, Braun T (2012) Content discovery in opportunistic content-centric networks. In: IEEE 37th conference on local computer networks workshops (LCN Workshops), 2012, pp 1044–1052, 22–25 Oct 2012

16. Lee J, Cho S, Kim D (2012) Device mobility management in content-centric networking. IEEE Commun Mag 50(12):28–34

17. Hermans F, Ngai E, Gunningberg P (2012) Global source mobility in the content-centric networking architecture. In: Proceedings of the 1st ACM workshop on emerging name-oriented mobile networking design—architecture, algorithms, and applications (NoM'12). ACM, New York, NY, USA, pp 13–18

18. Kim D, Kim J, Kim Y, Yoon H, Yeom I (2012) Mobility support in content centric networks. In: Proceedings of the second edition of the ICN workshop on Information-centric networking (ICN'12). ACM, New York, NY, USA, pp 13–18

19. Wang L, Waltari O, Kangasharju J (2013) MobiCCN: mobility support with greedy routing in content-centric networks. In: IEEE global communications conference (GLOBECOM), 2013, pp 2069–2075, 9–13 Dec 2013

20. Anastasiades C, Braun T, Siris VA (2014) Information-centric networking in mobile and opportunistic networks. In: LNCS on wireless networking for moving objects: models, approaches, techniques, protocols, architectures, tools, applications and services. Springer, Berlin, pp 14–30

21. Kim D, Kim J, Kim Y, Yoon H, Yeom I (2015) End-to-end mobility support in content centric networks. Int J Commun Syst 28(6):1151–1167

22. Rossi D, Rossini G (2011) Caching performance of content centric networks under multi-path routing (and more). Technical report, Telecom Paris Tech

23. Ghodsi A, Shenker S, Koponen T, Singla A, Raghavan B, Wilcox J (2011) Information-centric networking: seeing the forest for the trees. In: ACM workshop on hot topics in networks (HotNets)

24. Chai WK, He D, Psaras I, Pavlou G (2012) Cache "less for more" in information-centric networks. In: Bestak R, Kencl L, Li LE, Widmer J, Yin H (eds) Proceedings of the 11th international IFIP TC 6 conference on networking—volume Part I (IFIP'12). Springer, Berlin, Heidelberg, pp 27–40

25. Psaras I, Chai WK, Pavlou G (2012) Probabilistic in-network caching for information-centric networks. In: Proceedings of the second edition of the ICN workshop on Information-centric networking (ICN'12). ACM, New York, NY, USA, pp 55–60

26. Wang Y, Xu M, Feng Z (2013) Hop-based probabilistic caching for information-centric networks. In: IEEE global communications conference (GLOBECOM), 2013, pp 2102–2107, 9–13 Dec 2013
27. Ghodsi A, Koponen T, Rajahalme J, Sarolahti P, Shenker S (2011) Naming in content-oriented architectures. In: Proceedings of the ACM SIGCOMM workshop on information-centric networking (ICN'11). ACM, New York, NY, USA, pp 1–6
28. Smetters D, Golle P (2010) CCNx (Pre 1.0) Access control specifications. PARC Technical Report, 30 July 2010
29. Li Q, Zhang X, Zheng Q, Sandhu R, Fu X (2015) LIVE: lightweight integrity verification and content access control for named data networking. Inf Forensics Secur IEEE Trans 10(2): 308–320
30. Gasti P, Tsudik G, Uzun E, Zhang L (2013) DoS and DDoS in named data networking. In: 22nd international conference on computer communications and networks (ICCCN), 2013, pp 1–7, 30 July–2 Aug 2013
31. Mauri G, Raspadori R, Gerla M, Verticale G (2015) Exploiting information centric networking to build an attacker-controlled content delivery network. In: 14th annual mediterranean Ad Hoc networking workshop (MED-HOC-NET), 2015, pp 1–6, 17–18 June 2015
32. Ribeiro I, Rocha A, Albuquerque C, Guimaraes F (2014) On the possibility of mitigating content pollution in content-centric networking. In: IEEE 39th conference on local computer networks (LCN), 2014, pp 498–501, 8–11 Sept 2014
33. AbdAllah EG, Hassanein HS, Zulkernine M (2015) A survey of security attacks in information-centric networking. Commun Surv Tutorials IEEE 17(3):1441–1454 (third quarter)
34. Pentikousis K, Ohlman B, Davies E, Spirou S, Boggia G (2016) Information-centric networking: evaluation methodology, ICNRG, Internet-Draft, draft-irtf-icnrg-evaluation-methodology-03, Expires: 21 Apr 2016
35. Pentikousis K, Ohlman B, Corujo D, Boggia G, Tyson G, Davies E, Mahadevan P, Spirou S, Molinaro A, Gellert D, Eum S (2014) ICN baseline scenarios and evaluation methodology, ICNRG, Internet-Draft, draft-pentikousis-icn-scenarios-04, Expires: 16 Jan 2014

Printed in the United States
By Bookmasters